Of Population and Pollution - A Global Warming Primer

Of Population and Pollution - A Global Warming Primer

ONE MAN'S THOUGHTS ON WHAT WE NEED TO DO TO PREVENT HUMAN EXTINCTION FROM OVERPOPULATION AND GLOBAL WARMING

Arnold James Byron

© 2015 Arnold James Byron
All rights reserved.

ISBN-13: 9781511994408
ISBN-10: 1511994401
Library of Congress Control Number: 2015907681
CreateSpace Independent Publishing Platform
North Charleston, South Carolina

I want to thank my wife, Jean, for her understanding, and my granddaughters, Samantha and Sarah, for their help with the artwork for the cover pages and the chapter headings.

About the Artwork

THE FRONT COVER OF THIS book depicts a happy and healthy Earth after the population has been halved by responsible methods and Earth has been healed of any threat from global warming. The artwork for the front cover consists of a circle representing Earth; words denoting life, love, and happiness springing from the circle; seventy-five dots placed randomly within the circle; and a legend showing that each dot denotes fifty million people. The front cover depicts the largest number of people that I believe Earth can fruitfully sustain after we have stopped using fossil fuels and nuclear power as our sources of energy. This number is 3,750,000,000, one-half the current population. I am hopeful that Earth will be able to sustain this many people. The analysts at www.worldpopulationbalance.org claim a sustainable population should be only two billion people.

I often use the expression "specter of human extinction" in this book. As I wrote these words, I saw the need for an image that would personify the specter of human extinction. We need some sort of visual impact to help us keep a proper perspective on the impending ravages of overpopulation, global warming and nuclear power plant disruptions. The image of this specter is displayed on the back cover. I have also placed an icon of this characterization at the beginning of each chapter of this book as a reminder that overpopulation does exist and that the ravages of overpopulation are imminent.

This artwork is made up of a circle representing Earth, the sentence "Extinction is a word for nothing left to lose," and the sentence "There is nobody here." The sentence "There is nobody here" has been placed in the circle to show a face. The sentence "Extinction is a word for nothing left to lose" is used to make several circumscriptions around Earth to represent the greenhouse gasses surrounding the planet. It is also a paraphrasing of the lyric "Freedom's just another word for nothing left to lose" from the Kris Kristofferson / Fred Foster song "Me and Bobby McGee." This song has been covered numerous times since the original, and the "freedom" lyric may have been coined by Simon and Garfunkel. No matter the original source of the lyric or who sings the song, the point is the same: we hope that freedom is not lost to extinction.

I have used the definition of "specter" from *Funk and Wagnalls Standard College Dictionary*: "1. A ghost or apparition. 2. Anything of a fearful or horrible nature." I want you to know that even though my granddaughter and I tried very hard to make the image of the specter into a grim caricature, it came alive as a sad face. But don't be fooled. Grim or sad, the specter has its work to do.

I hope that my efforts to make the specter of extinction of the human race into a visual caricature will help you become more energized in your efforts to combat overpopulation and global warming. Don't give the specter any opportunity to manipulate minds and events to its purpose; rather, do all you can to help to prevent the specter from gaining its ultimate goal: the extinction of human life on Earth.

Love and happiness are the most positive attributes of people on Earth. These are depicted on the front cover. Live them to their fullest. Do so knowing that Earth's population can be regulated to a size that will easily promote the acts and arts of living. We must make the future into a time when love and happiness will shine, in nature and in spirit.

Table of Contents

About the Artwork vii
Introduction xiii
Prologue xv

Chapter 1   What Is Happening?
What Can We Do? 1
Chapter 2   Setting the Stage 10
Chapter 3   Problems and Solutions 15
Chapter 4   Responsibility versus Freedom 20
Chapter 5   Scientists 25
Chapter 6   A View from Wherever You
Choose to Worship 27
Chapter 7   What about the Oligarchs and
Corporations? 31
Chapter 8   What about the UN, the
World Court, and the IMF? 34
Chapter 9   What about the Military? 37
Chapter 10 What about the Common Folk? 39

Chapter 11	An Action Plan and the People to Work It.	41
Chapter 12	A Plan for the Nations.	50
Addendum 1	Reducing the Earth's Population	65
Addendum 2	Science and Science Deniers.	74
Addendum 3	Front-Yard Gardening	82
Addendum 4	Pope Francis's Encyclical on the Environment.	87
Addendum 5	It's All about the Living: A Letter to the Editor	98
Addendum 6	Are We at War?	101
Addendum 7	Controlling the Message	104
Essay 1	In a Nutshell	111
Essay 2	The Global Dilemma	113
Essay 3	Garbage to Electricity	120
Essay 4	Depolymerization Explained	126
Essay 5	Electricity by Steam Engine.	132
Essay 6	Countywide Cooperative	136
	About the Author.	145

Introduction

SUNLIGHT IS ONE OF THE three benign energy sources that provide power to Earth and its inhabitants; the other two are geothermal activity and gravity. Atomic energy is fearful and problematic. Even its waste products are radioactive and hard to contain safely.

Our planet has experienced (and is still experiencing) sunlight in three orders, which I like to call "ancient sunlight," "recent sunlight," and "instant sunlight." The fossil fuels we use today are ancient sunlight. The plants and animals living today are recent sunlight. The sunlight that touches our faces and creates the wind is instant sunlight. Instant sunlight is the energy supply for solar arrays and wind turbines.

The danger we face today is that we have put ourselves on the brink of nonexistence by overusing our supply of ancient sunlight. We must find a way to use our renewing supply of recent sunlight to augment and ultimately replace our current use of ancient sunlight while we strive to use instant sunlight to meet our energy needs.

Geothermal activity and gravity will be a part of the energy picture of the future, but to a lesser degree than solar applications. Atomic energy must be eliminated.

We must find solutions to three main problems: we must reduce the amount of carbon dioxide in the atmosphere, we must peacefully reduce the number of people on Earth in a responsible and humane way and we must discontinue the use of nuclear power. My hope is that readers will find workable solutions to these problems herein.

Prologue

FOR THE PAST SEVERAL YEARS, I have been watching as the world has been warming. You have also been watching. We all see a problem that is global in scope and so huge and complicated that we cannot seem to wrap our minds around it. What are we to do about overpopulation and global warming? We all want the problem to go away. The experts have not stated any quick and easy solutions. We, as common folk, lurch from news article to YouTube video, wondering what can be done. Let us be reminded that a sizeable chorus of voices is crying out that the endgame of global warming might very well be the extinction of the human race.

  Let me begin by saying that I am worried. I am worried that the human population has grown so large that it is using up all of the world's resources. There are over seven and a half billion people on the Earth. The time has come for humanity to begin the task of reducing the size of the population to a more sustainable number. I am using this book to lay out my ideas on how to do just that. I hope that my ideas are recognized as being good enough to do the job.

Finding the way to solve the problem is only half the game. The other half is convincing the people that there is indeed a problem and that they need to become a part of the solution. People are set in their ways. Often they need a psychological boost or a psychological bump to remind them of their responsibilities.

I have tried to make this primer as nontechnical and easy to read as I could. I want you to think about the future and whether you should change your attitude toward the problems of overpopulation and global warming. It is never easy to get involved, especially when one needs to change one's mind in order to find the courage to get involved. I hope that this book will help everyone who reads it to find a way to get involved and do more.

Over the years, I have spent some of my time looking at the problem and trying to figure out solutions, and I have developed a number of ideas. I think of these ideas as mere thoughts, but I often wonder, might these thoughts lead to solutions? I don't know. But these thoughts will do no one any good if no one sees them. With that in mind, I decided to put these thoughts into book form.

A book needs heroes. Who are the heroes of my book? Who will take on the issues of overpopulation and global warming? Because the problem is global, the heroes must also be global. I have come to the conclusion that two groups qualify as heroes: the religious leaders and the scientists of the world.

I chose religious leaders because religion is global, and common folk trust their religious leaders. Solving this problem will require a huge element of trust. I chose scientists because the problem is essentially a science problem. Science is the study of the natural phenomena of the Earth. Let me put it this way: God is global; nature is global; God and nature are intertwined. Religious leaders and scientists are the elements who must come together to provide the other factions of humanity with the impetus to actively work toward solutions.

The other factions of humanity include business and government. People trust religious leaders and scientists; they do not trust business and government. Relying only on CEOs and politicians to solve such a huge global problem will not be enough. Yet the wherewithal needed to solve the problems of global warming and overpopulation lies in the hands of business and government. Scientists and religious leaders can give direction. Politicians can tax the people and make needed changes in laws. Business leaders can provide organization, experience, expertise, workers, overhead, and the rest of the wherewithal necessary to get the job done.

The problem with trying to solve global problems is that the problems are just that: they are *global*. What I mean by that is that everything is too large and too diverse. People who live on one part of the globe are perceived to be different from those who live on another part of the globe; the same holds true for the cultures and religions that people embrace. Biases and prejudices have existed from the start of civilization. How will it be possible for everyone to work together? I suppose this is why we need to have religion and science at the fore: everyone will listen to and respect them.

We have come to the place where all my thoughts and plans, no matter how good they may be, can crash and burn—because if the world's leaders in science and religion cannot (or will not) work together to persuade the world's governments and business leaders to put aside any agendas that would continue global warming, then nothing will happen. If the leaders of religion and science choose not to be heroes, then the world will continue to move toward extinction. Business and government are the entities that hold the power of legislation and money. They must be convinced that it is time to give up any agenda that will not enhance efforts to combat overpopulation and global warming.

We need our heroes. Our need is for our heroes to step up, be ready and willing to be criticized, and be heroic. Think in terms of finite Earth, infinite population, and diminishing resources. Then step forward and lead.

THINKING FROM LEFT FIELD

This book is the telling of one man's thoughts about what we need to do to prevent human extinction by overpopulation and global warming. This book is not intended to be a mere regurgitation of ideas and information first developed by others. It is my hope that this book will be a psychological booster for anyone who is worried that there is no way that we can develop a workable relationship with the natural forces of the planet so that life on Earth can continue safely with fulfillment for all. I began thinking about writing this book in the winter of 2014 and 2015. Any storms that we experienced should have been colder and more troublesome than they were. Instead, the weather in Washington State was warmer than usual. Storms brought rain instead of snowpack; we had another record-breaking, global-warming year.

The trouble with trying to figure out what to do about overpopulation and global warming is that this is the first time that the human race has had to confront the problems on a global scale. Because the problems we face are unprecedented, the solutions may also have to be unprecedented.

My intent is to set forth a few ideas for solutions to the dilemma we face. To my mind, the ideas I set forth are doable; otherwise, I wouldn't be writing this book. An old cliché goes something like this: "Where did you find those ideas? Somewhere in left field?" We face unprecedented times. Don't give up on ideas just because they may seem to be weird or too hard to do.

As I noted in the "About the Artwork" section, I have created an image of what I call the *specter of extinction.* Many people find it easier to focus their thoughts (and to know that a problem exists) when they can see a representation of the problem. The problem we need to solve is a global problem. People worldwide need to be focused on the things that will save the planet and the humanity that lives here. People need to be aware that the possibility of human extinction is real. My hope is that whenever someone sees the specter of extinction, that person will raise an eyebrow and say, "Yes. I will become energized. I will do more. How can I help?"

Do not think of the specter of human extinction as the opposite of God, or of having any religious significance whatsoever. My intention is for you to see the specter as representing your fears, your reluctances, and your unwillingness to become involved in the work of healing the planet. It is my intention that whenever somebody sees the image of human extinction, that person will react positively toward the work that needs to be done.

I want people to fear the future only to the extent that they begin to see why they must be prudent when they consider the world that will exist for their family members who are born now and later. Someone who is prudent might be expected to be cautious, discerning, vigilant, well advised, and on guard, yet ready to get the job done. When one acts with prudence, one acts in a forthright, well-thought-out, and positive manner.

Consider this question: Am I being prudent? How you answer this question for yourself will be the measure of the extent of your activism in creating a sustainable population and in healing the planet. I believe that in writing this book, I am being prudent. I charge you with becoming involved and finding your own way of helping.

Of Population and Pollution - A Global Warming Primer

This book is about global warming, but it is also about overpopulation. I will assert in the pages ahead that global warming is a consequence of overpopulation. I will further assert that the problem of global warming will not be solved until the problem of overpopulation is solved. Most people refuse to even think about—let alone consider the consequences of—overpopulation. These people may think of me as being out in left field. But the reality is that Earth is finite in size, and we are intent on putting an ever-increasing (perhaps infinite) number of people on it.

We need a World War III, and I don't mean the kind fought with guns and other weaponry. Overpopulation and global warming are problems that affect the entire planet. The problems and the solutions must be worldwide in scope. We are used to the terminology: the war against hunger, the war against poverty, the war against drugs. We need to recognize a war against overpopulation and global warming. This ought to be called World War III. The armaments to be used in WWIII will be (1) ideas and solutions, (2) the people who will bring those ideas and solutions to fruition, and (3) the logistics and infrastructure that will be needed to complete the task. The struggle will be worldwide, and the war will last for two or three generations. But after the war is over, humanity will live on a planet with less pollution and more happiness than we are experiencing today.

DISCLAIMERS AND EXPLANATIONS

This section is included so that I can draw your attention to a few specific parts of the book where I think some additional information or explanation may be necessary.

In chapter 2, I assert that if we continue at the present rate, the population of the world will grow to a size where the planet is no longer able to provide for the needs of the population. What will happen then? Nothing like this has ever happened before. Because I have no idea what life would be like in such a world, the descriptions I use (in chapter 2 and addendum 1) to suggest how hard and difficult life would become come from my imagination. I have simply tried to describe the way life might be in the most frightening manner I can think of.

I have watched websites and YouTube videos that claim there will be a breakdown of the economic system, followed by the breakdown of the technological system, followed by the breakdown of nuclear power plants' operational capabilities. Is such a thing possible? My questions are, What will happen to the nuclear power plants when the technology systems become entirely stressed out? and, Will we be able to shut them down before they melt down? If the worst-case scenario happens, I believe that life on Earth will become completely untenable compared to the life we live now.

I want to thank the many forward thinkers who, for the past several years, have tried to draw our attention to this situation. They are scientists, media personalities, journalists—and even businesspeople and politicians.

They all deserve credit but are too numerous to list here. Look them up on the Internet; they are easy to find.

Chapter 8 has to do with the money that will be needed to pay for all of the expenses incurred in our effort to save the planet. A time will come when huge amounts of money will be required to pay for all of the work that will need to be done. Where will the money come from? Who has the money? What agreements will be made to make the money available? How will the money be used? And who will use it? These questions will all be answered when world leaders finally decide to act. There is no way to know how things will play out, because nothing has happened yet.

But my real fear is that overpopulation will overpower Earth's resources before we even begin to make decisions. We need to start finding all that money now so that we can get a head start on averting the worst-case scenario.

Chapter 11 tells who the players are and what they need to do. I hope that in the near future, these players will take up their roles and get on with the work of saving the planet. I have written the chapter in an odd sort of way. I want so very much to end the book on a positive note, with a win for all, that I have written the conclusion so that there is a win every step of the way.

These are the steps: (1) The religious leaders and the scientists form the core of the effort; (2) the politicians pass laws that enable the effort to go forward, including laws to limit childbirth to one child per couple; (3) corporations, oligarchs, and governments produce the necessary money; (4) the engineers and workers build the necessary infrastructure and grow food to feed the population; and (5) Earth is safe, and life is good.

Giving the book a positive ending is good for our psyches. I hope that reading a positive ending will inspire the reader to decide to work hard at solving the problem.

Addendum 1 is where I lay out my plan to reduce the population of the world by half. I don't say anything about forced sterilization. I do say that every male must be required to get a vasectomy after he has fathered one child. Is that forced sterilization? I say it is not, because forcing every male to get a vasectomy after fathering one child will be contributing to a common good and will be required of every male on Earth. If it is not required of every male on Earth, then it would be forced sterilization and would be evil. I think of forced sterilization as an attempt at genocide, or of reducing the population of certain groups of people.

I have been asked why my plan says that in the case of a divorce, the child should be given to the father unless a court orders differently. I say this because the father, after undergoing a forced vasectomy, will not have the ability to have another child. He has given up his rights for a noble purpose. As this program is developed by legislation and the courts, things such as visiting rights and parental obligations will be considered. This program of population control will be completely different from anything that has gone before. Get ready for a lot of confusion.

Addendum 3 is where I set forth my ideas about a concept that I call "front-yard gardening." The idea is to find a way to grow food for all the people while at the same time using farmland to grow crops that will be used to produce the energy that will replace the energy we currently get from burning fossil fuels. Naysayers will say, "Why bother? It's going to take energy to grow and distribute the food that's being grown in front-yard gardens."

I say, "People who live in the United States eat oranges grown in Chile." Any new methods we develop to grow and distribute food products will be more energy efficient than our current methods. I think that people are clever enough to invent and develop new methods to grow the food we eat: methods that are less expensive and more efficient than what we are doing now.

Hydroponics, aquaponics, and vertical greenhouses are methods that have been designed to produce more food per square foot than our current greenhouse configurations allow. These emerging methods can be developed and will help grow enough food for all. Reams of information on hydroponics, aquaponics, and vertical greenhouses can be found on the Internet.

Addendum 7 tells about a plan to build and operate a news media network of radio stations, television stations, and printed media. I found a neat picture graph at www.thinkprogress.org. It shows the differences between conservative and progressive programming in ten top radio markets in the United States. It shows that an unreasonable disparity exists. Conservative programming is 76 percent of the market, while progressive programing is at 24 percent of the market. See below.

Radio Market	**Conservative**	**Progressive**
San Francisco	69	31
Los Angeles	69	31
Dallas	100	0
Houston	100	0
Chicago	53	47

Detroit	60	40
Atlanta	96	4
Philadelphia	100	0
New York	53	47
Washington DC	65	35
Total	76.5%	23.5%

Why are Dallas, Houston, Atlanta and Philadelphia so out of balance? I should think it would be in the best interest of the United States of America, its people and its governance to insist on equality of programming everywhere in the country.

We need to reach out to the vast majority of people who don't seem to be concerned about overpopulation and global warming. I think this can be done, especially if we come together and work together as nonprofits—either corporations or cooperatives. If building and operating such a network is where your strengths lie, I can assure you that my short outline of the plan has barely scratched the surface of possibilities. You are welcome to become involved with and to work to develop this critical aspect of the future.

NOTES AND TIDBITS

The following section is included as a place where I can present a few thoughts that are pertinent but not included elsewhere in the book. As I wrote this book, I decided to arrange my thoughts into three distinct sections. The first section contains twelve chapters. This section establishes the problem, designates the actors who will work to solve the problem, and provides them with an action plan. The second section is made up of seven addenda written more or less in conjunction with the first section. I used the addenda to spell out and to give depth to my ideas. The third section is the essay section; it is made up of a series of essays I wrote starting ten or more years ago, when the problems associated with planetary warming began weighing on my conscience more heavily than they had before. Back then, I tried to share the essays and my thoughts with politicians, friends, and family, but to no avail. I share them now as a part of this book. Like the addenda, they spell out and give depth to some of my ideas.

What you will read in this book are the thoughts of one ordinary man about what we need to do to prevent the extinction of the human race by overpopulation and global warming. Three ideas describe the essence of the obligation we have to the human population and to the planet: prudence, prevention, and action. Keep these words in mind as you go forward to help in any way you can.

From time to time in this book, I call on experts to become the leaders we need. My definition of an expert is anyone who is trained and knowledgeable in any particular field or endeavor. If the need is for an expert in science, then a scientist or engineer may be the expert who needs to step forward. If the need is in law or business, then the expert who steps forward will likely be from those fields.

Speaking of fields and experts, let me say, "The farmer is outstanding in his field." Pun intended. As you read this book, you will see that my plan relies heavily on farmers growing crops to remove carbon dioxide from the atmosphere and master gardeners managing the gardens that will feed the people. I am confident that experts from every field will step forward.

I want to comment on my plan to reduce the number of people on Earth. If it is done in the way I describe, then the ratio of one population to another, or of one nation to another, will be the same after the population has been halved from what it is now.

We already have fuel sources that are made from recent sunlight. Biodiesel is made from oil that has been extracted from corn. This corn oil is a good substitute for diesel oil. Corn is also being used to make ethanol, a form of alcohol. Car engines can be modified to use alcohol instead of gasoline.

The gasoline and diesel fuel that can be made by what is known as "thermal depolymerization" will also be made of recent sunlight. It is my personal opinion that, when fully developed, the thermal depolymerization process will make diesel fuel and gasoline more efficiently, with better results, than is happening at this time for biodiesel and ethanol. (For more on this topic, see essay 4: "Depolymerization Explained.")

The knowledge and information that has stimulated my thinking about global warming and overpopulation has been gained from newspapers, magazine articles, books, radio, television, and Internet sources (including websites such as YouTube). I have been worried about and thinking about this global problem for at least two decades. My ideas have evolved over this period. Much of the information is merely remembered from browsing or reading in a casual manner. I have tried to find sources for these ideas and hope that I have not overlooked giving due credit for any citable information or ideas.

Some parts of this book are wholly my own ideas and inventions; I claim intellectual property rights for these. They are found in the following places: chapter 12, "A Plan for the Nations"; addendum 1, "Reducing Earth's

Population"; addendum 3, "Front-Yard Gardening"; addendum 7, "Controlling the Message"; essay 5, "Electricity by Steam Engine"; and essay 6, "Countywide Cooperative."

Here is a fictitious story that depicts a scenario that, sadly, is too much to expect. Is it too much to expect that one morning a general from one military force should see the specter of human extinction and arrange a communication with a general from the enemy forces and say, "From general to general, let's study this possibility of world extinction and see if it would be better for us to join forces in a war against extinction instead of trying to kill one another"?

Here's an idea: improvise a parlor game using the format below. How many different scenarios leading to a final consequence can you think of?

- Global warming is a consequence of overpopulation →
- Drought in the western United States is a consequence of global warming, which is a consequence of overpopulation →
- Wildfires are a consequence of drought in the western United States, which is a consequence of

global warming, which is a consequence of overpopulation →
- Loss of property and wildlife is a consequence of wildfires, which is a consequence of drought in the western United States, which is a consequence of global warming, which is a consequence of overpopulation →
- Etc.

What is the final outcome of the consequences of the stress we put on the planet's resources?

I was in elementary school (likely fifth or sixth grade) when I learned that the Sahara Desert was at one time lush and green and full of life. I wasn't much older than that when I learned that the fabled cedars of Lebanon had all but disappeared. Do these phenomena have anything in common? I believe they do. And that "anything" is overpopulation. It is what I would call "regional overpopulation." The "Human Overpopulation" section of Wikipedia says that overpopulation occurs in a defined area when the number of people exceeds the "carrying capacity" of the region occupied by those people. Carrying capacity is all about sustainability. If a region cannot sustain the population of that region, then the population is doomed. People must die off, leave, or have fewer children.

I suspect that the Sahara Desert became what it is today by virtue of too many people needing to raise too many sheep and goats for sustenance. As for the cedars of Lebanon, I would venture the guess that too many people living in the region needed too much timber for temples and houses and whatever else is made from wood. The cedars of Lebanon might be standing today but for the overuse that was necessary to sustain the overpopulation.

The overpopulation we are experiencing now is more than regional. Let me put it this way: *regional is now global*. Everything that happens to the population of a region will happen to the population of the whole Earth. We are already experiencing the distress. We are coming to the point where people must die off, leave, or have fewer children.

We cannot simply die off. Life is for the living. We cannot move. There is no other Earth close to us. But we can have fewer children.

The words "climate change" will be found in this paragraph and in only a few other places in the book. My thought is that using the words "climate change" as a euphemism for global warming dumbs down the whole concept that greenhouse gasses are warming the planet. The greenhouse gasses are mostly carbon dioxide, methane, and water vapor. The term "climate change"

is relevant only because the excess water vapor plays a significant role in the severity of storms and other climatic activity. Carbon dioxide is well known as a greenhouse gas, but it is methane that needs to be given more exposure.

The term "climate change" tells someone that a problem exists, but the listener does not delve more deeply in order to see the whole problem. The danger posed by methane is so hugely diabolical that we all must learn about it. When the words "climate change" are used, people think they have all the information they need, but they need to know more. The reality is we all need to know more.

As you read this book, you will see that I believe the time has come to start removing carbon dioxide from the atmosphere. You will also see that I favor thermal depolymerization as the best way to do this. When thinking about how many thermal depolymerization plants will be needed, I think: lots of them. When thinking about how large each of these plants should be, I think: only large enough so that household garbage and the other feedstock can be brought to the plant using only the minimum amount of energy. Each thermal depolymerization plant will be sized differently, but the general rule will be, smaller is better.

So I have been asked the following questions. Why is bigger not better, especially in terms of the business concept known as "economies of scale?" Why waste money on building several small units when it may be cheaper to build one large unit? Wouldn't it be better if fewer people could run a large plant rather than hiring a larger number of people to run several smaller plants? Think of all the money you could save.

My answer is, "I see a future where everything is local." When generating electricity, we should generate enough for local use and then sell the rest to the power grid. Thermal depolymerization that will change household garbage and other carbonaceous (i.e., carbon-rich) materials into electricity and petroleum oil should be done locally so that the energy we make is not wasted on transporting either raw material or finished product over unreasonable distances. I recommend that we base our solutions on local endeavors. I further recommend that the solutions be based on the simplest technology available. Solar panels, windmills, and depolymerization plants are low tech and local.

When thinking about removing carbon dioxide from the atmosphere, we should be thinking as if we are at war. When fighting a war, we are not so much concerned about the amount of money we spend; it is all about how much firepower we have. What we must do will require that we act in ways that might not be considered good business practices in the corporate world. Smaller is better.

Scientists have known for years that the way to improve flue gas emissions on coal-fired power plants is to install sulfur dioxide flue scrubbers. This makes sense, and it helps reduce greenhouse gasses. I like the notion of scrubbing to make things clean. This book is about scrubbing carbon dioxide from the atmosphere. I say we can scrub the atmosphere by simply growing plants.

How can this be? Let me explain. When plants are growing, they use carbon dioxide. We can harvest the plants and then use the thermal depolymerization process to turn them into a light crude oil that is much like the petroleum oil we pump from the ground. Then we can turn the crude oil into gasoline, diesel oil, regular oil, and electricity. What's left at the end of the process is carbon that we have scrubbed from the atmosphere. This is what thermal depolymerization is all about.

On May 11, 1935, President Franklyn Delano Roosevelt signed an executive order to establish the Rural Electrification Administration (REA) so that electricity could be brought to farms and rural areas in the United States. Supplying every home with electricity had become problematic and because of the Great Depression the established electrical utility companies were unable to get the job done on their own. The help that government was able to give resulted in expanding the economy, creating new jobs in many sectors of the economy and in making life better for rural dwellers and city dwellers alike.

That was the past. What about today? Should we expect government to step in and help to build the infrastructure of solar (includes wind), geothermal and gravitational applications that will produce the energy of the future? Should we expect government to step in and help to build the infrastructure of pyrolysis/depolymerization plants to scrub carbon dioxide from the atmosphere? The REA was a success story. Can we have a new success story?

The information given in this tidbit is common history from eighty years ago and doesn't need a citation. But I happened to stumble onto a website called great achievements dot org. It is my hope that humanity will have a sustainable future in which to recognize the great achievements of the future.

I heard a discussion on the radio the other day about using specialized greenhouses to grow more food in less space. Vertical greenhouses can be built several stories high. The containers of plants are put on shelves placed one above the other, all the way up. The footprint of a vertical greenhouse is very small compared to the footprint of a regular greenhouse.

Now I want you to do a mental exercise. Think about a man-made pond large enough to grow fish for harvesting. Then think about a vertical greenhouse, several stories high, built on top of that pond. Now think about how the enriched wastewater from the fishpond can be pumped up to feed and water the plants in the vertical greenhouse. You are thinking about aquaponics. In the near future, aquaponics may become one of the best ways to provide the greatest amount of food. The best way to learn about aquaponics is simply to look it up on the Internet.

The people on the radio strongly emphasized that new greenhouse technology could help grow enough food to allow the population to continue to increase. Developments in new greenhouse technology will no doubt be a boon to the people of the world. But what if simply growing more food is not enough to sustain the growing population? Over seven billion people currently live on Earth, and we are expected to grow by a billion more every generation. At some time in the future, the population will outgrow the Earth's resources. I think humanity needs to reduce its own population to ensure that this battle against extinction can be won. Then regulate the size of the population, forevermore.

What do I want for the future of Earth? I want a population of a size that lives within the constraints of the Earth's resources. I want the thermal depolymerization process to be the way to produce gasoline and diesel fuel. I want the production of electricity to come from windmills and solar panels and from geothermal, tidal, and wave action.

I am often asked the following questions about the depolymerization process. If you make gasoline by the depolymerization method and then use the gasoline when you drive your car, aren't you putting the carbon dioxide back into the atmosphere? What good have you done? I will answer these questions by sharing a fictitious conversation between two fictitious friends with common first names, Mike and Bill.

Mike: Bill? You're a smart guy. I have a question I want to ask you.

Bill: Go for it.

Mike: I've recently heard about this process called thermal depolymerization. I'm told you can make gasoline by using this process.

Bill: Yeah. That's what I understand. I ran across that process in a book entitled *Of Population and Pollution*. What's the question? I hope I can answer it.

Mike: Well. Here's my first question. Is getting this new kind of gasoline different from what they do in a refinery?

Bill: No. Not at all. It doesn't make any difference whether you are refining regular petroleum oil or the oil you get from this depolymerization method. You're still gonna get the same products: gasoline, diesel fuel, and all the rest. The only difference is that the refinery uses petroleum oil that has been pumped out of the ground, whereas the depolymerization process changes garbage, old tires, and plant matter into a kind of petroleum oil. They are both refined in the same way.

Mike: Then how are they different? And besides, if you drive your car using gasoline made from garbage and old tires, doesn't that put carbon dioxide into the atmosphere? I mean, we are told that we have to stop using fossil fuel because burning fossil fuel puts carbon dioxide into the atmosphere, so how can we justify going to a process that does the same thing? It sounds like a bunch of hooey to me.

Bill: Let me explain. But be patient. I'm gonna start by saying something completely out in left field. The difference between getting gasoline from fossil fuel and getting it from a depolymerization method is governed by how many solar panels and windmills we build.

Mike: Man, you are getting crazy. But I promise. I will be patient.

Bill: OK. Next is the gasoline itself. At the present time, we get our liquid fuels (gasoline, diesel fuel, and oil) from fossil fuel. Cars, trucks, and trains depend on liquid fuel. Will we ever be able to power trucks and trains on electricity from solar panels? I don't think so. Batteries that will store and release that much power haven't been invented.

Mike: Okaaay? That makes sense, but where are you going with this?

Bill: Well. You just said we have to stop using fossil fuel, but we need to have gasoline and diesel fuel. If we aren't going to make gasoline from fossil fuel, then we will have to find another way. The depolymerization process is the only way I know of to make gasoline and diesel fuel from something other than fossil fuel.

Mike: But that's still dumb. You're still putting carbon dioxide into the atmosphere even if the gasoline is made from old tires.

Bill: You are right, of course. But what will happen when we stop making gasoline out of fossil fuel?

Mike: I don't know what will happen. We will still need gasoline. You tell me.

Bill: You know that three kinds of products come out of a refinery: gaseous, liquid, and solid. The same is true for depolymerization products.

Mike: Okaaay?

Bill: Let's talk about the products from depolymerization. Depolymerization will change household garbage, vegetation, and old tires into the following products: natural gas, gasoline, diesel fuel, fertilizer, and carbon. The fertilizer and carbon are not burned and do not put carbon dioxide back into the atmosphere. But the important consideration is, where did the carbon come from? The paper, cloth, tires, wood and vegetation that make up the household garbage grew from carbon dioxide that came from fossil fuel, over the past two hundred years. Plastics are also made from oil that has been extracted from fossil fuel.

Mike: OK. I get that. The depolymerization process removes fossil fuel from the atmosphere. But you still haven't answered my question.

Bill: Well, just think about it. Using gasoline made from fossil fuel puts carbon dioxide into the atmosphere. Using gasoline made by the depolymerization process also puts carbon dioxide into the atmosphere, but at the same time it takes carbon dioxide—which came from fossil fuel—out of the atmosphere.

Mike: Now I get it. If you are doing both at the same time, the carbon dioxide in the atmosphere will increase. But if you are not doing fossil fuel and only doing depolymerization, then you will decrease the amount of carbon dioxide in the atmosphere even as you are adding some.

Bill: Yes, that's how it will work, and it will be a slow process. We will need hundreds of thousands of depolymerization facilities and more than one generation to remove as much carbon dioxide as we need to.

Mike: I see what you mean about the number of solar panels and windmills we put up. The more energy we get from alternative sources, the more we can get away from using fossil fuel. This is good because we will have depolymerization facilities to make the gasoline not being made from fossil fuels. We will continue to have the gasoline and diesel fuel we need, and we will clean up the atmosphere as well.

Bill: I hope I have answered your questions.

Mike: You have. For sure. I guess I know what we will be doing in the near future.

Bill: Oh? What will we be doing?

Mike: Building an infrastructure of solar panels and windmills.

Bill: I totally agree. That is what we need to be doing.

Mike: I guess I know what else we will be doing in the future.

Bill: Oh. What more will we be doing?

Mike: Building depolymerization facilities. Hopefully?

Bill: Yes. Hopefully.

This question is about the future and the present. What if humanity did not have to worry about overpopulation and global warming? If that were the case, then humanity would continue to use Earth's supply of fossil fuels until they are gone. This begs the question: What will humanity do after all the fossil fuel is gone?

If by that time a new energy infrastructure has not already been built, then humanity will be caught with its collective pants down, so to speak. Humanity will have to, at that future time, start building an infrastructure of windmills, solar panels, and depolymerization plants, plus the harvesting of geothermal, tidal, and wave activity in order to manufacture the energy people will need. Will it be too late then to start? I think so. My guess is that it will take twenty or more years to build an infrastructure that will provide new sources of energy. I think you will agree that we should start now, in our time. This is the prudent thing to do.

Here is a weird but thoughtful consideration. If we cut the population by half, then half of our houses will be empty. The owners of these empty houses should not have to bear the brunt of a decision to reduce the population; rather, the owners will have to be paid for the houses. But that's not the problem. The houses should not be left to rot away. The houses should be carefully dismantled and put into storage where the building materials will be kept dry.

I worry about the gypsum board. Gypsum board is the interior wallboard that is made of a white, chalklike substance. Gypsum is mined. It is not made. It is one of

Earth's resources. Gypsum cannot be mined forever. Earth will run out of it.

The problem with gypsum board is that it will be ruined if it gets wet. And then it won't be good for anything. Saving the gypsum and other building materials will become a priority in the future. The building materials from those houses will have value and will be able to be used at some future time.

What if the population of Earth is not reduced in a nonviolent, humanitarian way? The obvious answer: the population will be reduced by the chaos that will result from unmitigated overpopulation. Then all bets are off. Will all of the empty houses be cared for? The houses will simply go unused and rot away. What a waste. That would be very sad.

Let me tell you something that you absolutely do not want to hear. We can all agree that solar panels and windmills will be the energy of the future with the exception that we will need some gasoline and diesel fuel for trucks, trains and ships. Large vehicles are heavy and will require a lot of power. Battery power may not be enough. What about airplanes? If we cannot make enough fuel for airplanes, then airplane travel will become a thing of the past. But that shouldn't be much of a problem because the only thing we will have to give up is convenience. I told you that you didn't want to hear this.

If the invention of strong batteries should become a reality, then petroleum products from depolymerization could provide enough liquid jet fuel so that humanity can keep some air travel. I'm just saying that if humanity can keep air travel then aficionados may still be able to fight over the differences between travel by air and travel by land or sea.

Our lives are going to undergo lots of changes when humanity starts its war against overpopulation. The many changes that will come from reducing the population will become part and parcel of everyday life. There will be changes in energy production, energy use, food production, jobs, and more. One of the many changes will be about fewer changes. Diaper changes are what I mean. Reducing the population will mean fewer children, fewer diapers, fewer toys, fewer bicycles, fewer everything.

This translates to less manufacturing and less employment in every phase of the manufacturing sector. Adjustments will have to be made because humanity won't need as many schools, as many teachers, as many stores, as many doctors, as many nurses, as much of everything until the time of population reduction has been completed. Nations' economies will have to be closely controlled, with an emphasis on helping everyone stay on an even keel for up to two or three generations. Until this huge problem has been resolved everyone will have to be eager and willing to use their resources for the betterment of all humanity.

Sometime in the future when the people living today will have peacefully ended their life-spans, when the current supply of manufactured goods will have been used up and replaced by new things, and when all of the societal and manufacturing functions will have been recalibrated to the newly sized population, life will go on. How pleasant a thought is that? By the way, humanity will still need diapers. It's all about the living. It's all about the future.

My plan to reduce the population is simple. One child will be raised to adulthood by two people. It requires that every male undergo a vasectomy after fathering one child and that any female who becomes pregnant after having given birth to one child will promise to carry that pregnancy to term and give the child up for adoption to a childless couple.

There are many reasons why a couple may be childless, and each couple deserves to be able to participate in raising the children of the world. I believe that this reasoning also applies to nations where gay and lesbian couples are entitled to be married. If the goal is to cut the population in half, then every couple must participate.

Here is a useful thought about the past and the future. We know and accept that Earth began as a fiery inhospitable place. Earth cooled; plants and animals grew and covered Earth. It was at this time when Earth was lush and bounteous—filled with resources, a veritable Garden of Eden—that humans came into being. Since then the human race has grown to nearly seven and a half billion people. Lush and bounteous has become meager and wanting. Now, at this time, Earth is unable to supply all of the resources needed for sustainability.

The decline from bounteous to wanting is the consequence of overpopulation. Since the beginning of the human race nobody has had to think about overpopulation, until today. Now, the generations living today have to decide: Do we continue increasing, stay the same or begin decreasing the population? If we make a decision and act on it we are doing population control. Nobody wants to think about population control; much less do population control.

The math of population control is easy. More than two children equals increase; two equals stay the same; and less than two equals decrease. We need to start by decreasing the population in a humane, nonviolent, and non-eugenic manner. See addendum 1 for my ideas on doing this.

Humanity started with only a few humans in a lush and bountiful place. It has grown to billions; is using up Earth's resources, and may not be able to sustain itself. There has never been a need for restrictions on population growth. No good thing can last forever. It is the generations living today who have to develop, accept and apply the rules for population control.

The generations living today must accept the past and work toward the future. Whether the human population came about because of happenstance or providence is not a part of this discussion. This is about the destiny of the human race. It appears to me that population control has been the destiny of humanity from the beginning.

People tend to remove themselves from life's contingencies by losing themselves in entertainment such as music, movies, and sports. Please consider the following: it really makes no difference which team wins the Super Bowl or the World Cup; life goes on. But if the specter of human extinction wins, life ends. Which is the more important to get excited about: entertainment or life? Live your life to the fullest. Choose living. Become involved in saving humanity.

Here are some questions. What countries used five-year plans in their governance? For how many years should we be planning ahead as we consider the future of humanity? Is five years enough? Josef Stalin used five-year plans when he was dictator of the USSR and China uses five-year plans, even today. I do not hear news reports about how other countries are planning ahead. It seems that everyone is intent on what is happening in the moment.

Scientists say that the Sun will explode in two billion years. Does this mean that we should plan for the continuation of humanity for one billion years, or one million years, or one thousand years, or one hundred years? We need a plan. Reduce the population to a size that can be sustained by Earth's resources. Use thermal depolymerization, solar, geothermal and tidal sources for energy. Get rid of everything nuclear. This plan will last for a very long time.

CHAPTER 1

What Is Happening? What Can We Do?

AT SOME POINT IN THE future, a number of very serious events are going to play out. The people who inhabit Earth think of our planet as being gigantic. It may be large, but it is not infinitely so. Just as a box can be filled with only so many books, the planet can be filled with only so many people. We are filling the planet with people and will soon have to stop. Why? Because Earth is finite. It can only give the resources it has to give. Earth cannot sustain a population that grows without end. Earth's resources will come under ever-growing stresses until it can supply no more. Then there will be consequences—and these consequences could be huge. What will happen if we are not able to control the consequences? Chaos will reign.

What will the consequences of overpopulation be? Many scientists and other writers predict chaotic times, with little hope that stability will return until after a significant decrease in the population has come about through death and destruction. Some writers even go so far as to predict that the final consequence of overpopulation will be the total extinction of the human race.

The first manifestation of the impending chaos is already happening: this is the warming of the planet brought on by too many people using fossil fuels for energy. It is important to remember that warming the planet means warming the air, the land, and the oceans. The planet is warming because gasses that cause global warming, called greenhouse gasses, are being released into the atmosphere at rates that exceed what is natural. Carbon dioxide, methane, and plain old water vapor are the gasses that are being released in the largest amounts. Warming occurs because these gasses act like a transparent blanket spread over the Earth. During the day, the light and heat from the sun can easily enter the atmosphere. But the blanket will not let enough of the heat escape at night to maintain normality, so the planet heats up.

According to the US Environmental Protection Agency (EPA), carbon dioxide accounts for about 76 percent of all greenhouse emissions globally. Excess carbon dioxide has to go someplace, such as into the oceans and into the air. Nature uses carbon dioxide as a building block in trees and other plants. The effects of carbon dioxide depend on where it is found:

- Absorbed in the ocean, carbon dioxide can become carbonic acid, which kills life.
- In the air, it causes global warming.
- The carbon from carbon dioxide is integrated into trees and plants as they grow. The oxygen goes into the air, which is a good thing.

Much of the energy we use comes from trees and plants that lived millions of years ago. When fossil fuel burns, the carbon dioxide that had been broken apart by photosynthesis recombines with oxygen and becomes whole again. Once formed, carbon dioxide can remain unchanged for up to a hundred years.

Again, from the EPA, methane gas makes up about 16 percent of all global greenhouse emissions, but it has a warming potential that is about twenty-five times that of carbon dioxide. Methane gas is found in the atmosphere and in the ground. In the ground, certain bacteria will change the methane gas into carbon dioxide and water. In the atmosphere, methane gas will interact with certain oxygen molecules called free radicals and change into carbon dioxide and water. In any case, methane, which is a greenhouse gas, changes into carbon dioxide and water, which are both greenhouse gasses. Methane is bad stuff. We need to stop making methane gas.

Where does methane gas come from? Methane is made when we allow things to decompose without oxygen, such as garbage piles and cattle feedlots, or it comes from animals and people as food is being digested. Also, please note that the extraction and transportation of coal, natural gas, and petroleum are also significant sources. We can stop making methane gas by not dumping our garbage into huge piles, by ending our use of cattle feedlots for meat production, by not using fossil fuels, and by humanely reducing the number of people who inhabit Earth.

Methane has its own special secret. When methane gas is mixed with water at freezing temperatures, it can be frozen into ice crystals. Frozen in ice, methane is safe. But once heat is added, the ice crystals will melt, and the methane will be released. Huge deposits of frozen methane crystals and methane hydrate crystals are found in the tundra near the North Pole and in deposits under the ocean floor around the world. Many scientists are worried that if the tundra and the oceans get too warm and the methane crystals begin to melt, the release of methane will be of such a magnitude that global warming will become unstoppable.

Water vapor is the moisture that evaporates from land, lakes, and oceans. Too much water vapor in the atmosphere makes for more rain, more snow, and stronger storms. This is the "climate change" we keep hearing about. It seems to me that if we reduce the carbon and methane, the water vapor will simply return to normal levels.

Whenever I think about the problems that greenhouse gasses cause, I get the feeling that we must ready ourselves to take on global warming with the strongest fight we can muster. But I find that the majority of citizens—even the majority of academic and government leaders—don't seem to share the fear of the future that I have. When I search the Internet for information about what is happening, I find conflicting statements from various groups:

1. A very large number of solid, scientifically oriented scholars and government figures from around the world say that global warming is real, that it is created by humans, and that people living on Earth must work together to solve the problem.
2. A smaller number of equally solid and respected scientists and government officials warn that overpopulation, huge methane releases, and the meltdown of nuclear energy plants will cause the total annihilation of the human race.
3. A huge number (in the center of the issue) say that the planet is capable of maintaining itself, that carbon and methane emissions have started to slow down a bit, that population growth will not become a problem for another hundred years, and that we need not worry.
4. A very small but vocal fringe number of deniers say that global warming is not happening.

These groups should be named as follows: the groups who say the worst is yet to come, the group that says that everything will be all right, and the group that doesn't seem to care. Which group are you in?

I have learned that whenever I am required to do an unsavory job—or any job, for that matter—it is best to prepare for the worst possible scenario. Then, if that scenario doesn't unfold, I have a greater chance of succeeding, and the job is easier and produces better results. This makes good sense, and I intend to use this book to emphasize that we must know the worst possible scenario to know how and where to take our fight against overpopulation, global warming, and the specter of possible human extinction.

Read on. I have laid out a plan of action. I have made it clear that people need to take action, and I have specified the actions that they must take. I am in the group that believes the worst is yet to come. I pray that you will help resolve the worst-case scenario.

We are already experiencing an extremely large population, with over seven billion people vying for available resources and billions more on the way. But there is hope. The alternative to waiting for the population to outpace the Earth's resources is to prevent overpopulation. If we take control and work to reduce the size of our population while the planet can still provide sustenance for all people, we can prevent the chaos that will otherwise occur. It is up to us. If we already have children, we will have to accept the responsibility of instituting population control as outlined in addendum 1: "Reducing Earth's Population." With this system, two people will have only one child.

Now I want to draw your attention to seven conditions or elements of life:

1. Global warming is a consequence of having too many people on Earth.
2. Overpopulation is a condition that Earth is experiencing for the first time.
3. Our social structure dictates how we relate to one another.
4. Our economic system dictates how we can get the things that we need.
5. We currently rely on a technology system that is powered by electricity.
6. We need an electrical infrastructure that is powered by solar and wind arrays.
7. Nuclear power plants may not operate safely when they are placed under the worst conditions of neglect.

These elements are like characters in a movie production. Depending on the scenario they create, the writers and directors can make these characters into scary and macabre apparitions or into happy and resourceful images. I have outlined a few scenarios below in which we, as organizers and doers, can facilitate the best ending.

Scenario 1
The most disastrous scenario is one in which overpopulation and global warming create so much stress on available resources that the economic system fails, followed by the failure of the social system. At this point,

the technology system may not be able to sustain itself. The safe operation of nuclear power plants depends on a functioning technology system. In this scenario, the technology system fails—the power plants melt down, releasing huge amounts of radiation. All human life is extinguished. *This is the worst-case scenario.*

SCENARIO 2

In this scenario, we decide to dismantle all of the nuclear power plants in the world so that no nuclear power plants will remain to emit radiation. Then, when the economic and social structures fail, the chaos that follows kills off only a large part of humanity, not our entire species. Following this time of chaos, humanity rebuilds and continues to live on Earth.

SCENARIO 3

In this scenario, the people of the world decide to remove the excess carbon and methane from the atmosphere, reduce the population in a humane (i.e., nonviolent) manner, dismantle the nuclear power plants, and build an infrastructure of thermal depolymerization plants to provide gasoline and diesel fuel as well as an infrastructure of electrical generation by solar and wind sources. Bringing these decisions to fruition results in saving the planet and its population from chaos and annihilation. Controlling the size of the world's population, as well as the amount of carbon we put into the atmosphere, provides humanity with a sustainable lifestyle of safety and happiness that is free of stress and worry.

We have said for years that we must stop global warming. I submit that we must do more than talk. We must take action. We must be prudent. Being prudent means we must make every effort to resolve scenario 1 and instead to leave scenario 3 as our legacy. The specters of death, destruction, annihilation, and extinction must not have the last dance. I give you the following as watchwords: *We solve the worst-case scenario. We know humanity and the planet are safe.*

CHAPTER 2

Setting the Stage

A "PRIMER" (RHYMES WITH "SIMMER") is generally a book that is used in elementary school or a book that contains basic instructions. I have chosen to call this book a global warming "primer" because it is intended to provide basic instructions that, if they are followed, will stop global warming and save the planet from destruction. But first, let's go back to elementary school.

When I was in elementary school—in the late forties and early fifties—my classmates and I learned that coal and petroleum came from huge ferns and other plants that lived on Earth millions of years ago. We were taught that when plants grow, they take in carbon dioxide and give off oxygen. As it turns out, there was an abundance of carbon dioxide in the atmosphere in prehistoric times, which caused the ferns and other plants to grow very large. As the plants died and fell, they were subjected to heat and pressure, which in turn changed the plant life to coal and petroleum.

We also learned that the oxygen given off by the plants and ferns enabled the animal kingdom to emerge. We humans likely would not be here if God, or nature if you wish, had not provided plant life to sequester, or store, the carbon. Now we are releasing that carbon dioxide whenever we burn fossil fuels.

At the present time in our journey on Earth, a great deal of disagreement exists between those humans who support the notion of an omnipotent and omnipresent God who created everything on Earth and those humans who rely on the notion that the forces that act on the Earth are natural. Now is the time in our mutual human journey for all of us to ease up on our own personal beliefs and to concentrate on working together. When reading this book, readers are welcome to substitute the word "God" in every instance where I use the word "nature."

The human race started with a few people. Now there are too many. Overpopulation has outpaced nature and is stripping the Earth of all of the riches nature has provided. Thinking of the future makes me feel sad because the choices we will have to make will be wrenching. All people—everywhere, worldwide—will be affected. Everyone will need to know and understand the magnitude of the changes that overpopulation will bring. Every person on Earth will need to understand that if we do nothing to reduce the human population of the world, then the ultimate consequence may well be the extinction of all life on Earth.

The history of animals indicates that whenever the population of a species exceeds its resources, then starvation, disease, pestilence, and other negative forces cause that population to diminish to a sustainable size. The species then will be able to flourish once again. The future may not work out in the same way for the human race because other factors, such as nuclear power plants, are at play. When unabated human overpopulation reaches a critical mass, it will not be starvation and disease that will do the killing, but nuclear reactors. How will this play out?

In order to live together, humans have tried to form societies and rules of behavior to guide them. People within their own families and other groups—such as cities, states, and nations—depend on one another for sustenance and protection. The rules humans make to guide their behavior are always in flux and are tenuous even during the most stable of times. What will happen when unabated overpopulation runs its course?

When overpopulation reaches a critical mass, our economic systems will begin to fail. Stock markets will go out of business; banks will close. Governments—local, state, and national—will become overstressed and dysfunctional. There will be no money to exchange for goods and services. It will be difficult (or even impossible) to get the things we need to live. Our social systems will fail. Ultimately, there will be no government except the government imposed by the strongest and cruelest among us. Hardships will abound. Hunger and deprivation will prevail. People will die by the millions, and daily life will be torture. We will not be able to sustain the old ways of living.

In these circumstances, if overpopulation is the only factor, people will die off to a point where groups will become small and isolated enough for them to begin anew. Humanity will be able to move forward again. But another specter will be lurking in the wings.

When we take the nuclear power plants into consideration, we can easily come up with another scenario. What humanity needs to realize is that after the economic and social systems fail, we will still have the huge system of technology that we depend on for communications, computerization, and the operation of nuclear power plants. What will happen if the technological system fails? There is a substantial likelihood that nuclear power plants will begin to melt down and emit radiation. I do not know if an exact science exists to explain the horror that would result from a meltdown of all of the world's nuclear power plants, but it appears that some scientists believe there would be more than enough radiation to annihilate the human race. Should anything like this happen, it may take more than a hundred thousand years for Earth to heal itself and allow life as we know it to reappear.

This doesn't have to happen as described. It doesn't have to happen at all. The next chapter starts the "instructions" part of this primer. You are not going to like what I have to say, but I have come to the conclusion that if humanity does not follow these instructions, it will contribute to the possibility that the chaos described above will prevail. I do not want my grandchildren, great-grandchildren, and future generations to have to live in the chaos I have described. I will fight to end population growth, end global warming, and keep our world intact.

CHAPTER 3

Problems and Solutions

THE PROBLEMS INCLUDE (1) OVERPOPULATION, (2) too much carbon dioxide in the atmosphere, (3) the possibility that all of our nuclear power stations could melt down simultaneously, and (4) the need to build an infrastructure of solar and wind power to generate our electrical needs. Be cognizant that these are global problems, and individual societies around the globe may face distinct challenges in making the solutions work. The hardest part will be getting everyone to work together (and to be willing to make changes in their lives) so that the solutions will work for everyone. Two or three generations of people will be affected because it will take fifty or more years for enough carbon to be removed to allow the atmosphere to cool and for the population to be reduced enough so that the planet can again be a safe and sustainable place to live.

1. Overpopulation

The solution to overpopulation is not simple, but it is achievable. Read addendum 1, "Reducing Earth's Population," to find out how humanity can reduce the number of people on Earth. We will have to institute population control worldwide for two or three generations. Can this be done voluntarily, or will it take government mandates before people will submit to having this done? I believe that governments will have to pass laws and regulations to force people into doing something that they should do voluntarily. People need to be shoved, cajoled, and forced into doing things they would not do otherwise.

2. Too Much Carbon in the Atmosphere

The solution for this problem is to build an infrastructure consisting of hundreds of thousands of "pyrolysis" facilities throughout the world. Pyrolysis, also called depolymerization, is a system that will change carbonaceous materials into petroleum products and electricity. Please see essay 4, "Depolymerization Explained," for a comprehensive explanation of this process. A simple way to visualize a thermal depolymerization plant is to think about a mini-refinery. A mini-refinery will use depolymerization to change carbonaceous materials such as old tires, paper, plastic and field crops into petroleum products in the same way that a large refinery changes petroleum oil into petroleum products. They will be small but local and numerous.

How will we build such an infrastructure? A mini-refinery is comprised of various components that must be manufactured and then assembled. Private corporations will be the likely manufacturers. It doesn't make any difference whether the money to do the manufacturing comes from private or government sources, but the components and other hardware must be available before anything can be done.

If corporations, oligarchs, or governments will not step up and manufacture the components and assemblies that are needed for depolymerization plants to take shape, then there is not much we can do.

Then again, the manufacturing process may begin, but money may not be available to build the plants. Please see essay 6, "Countywide Corporation," which outlines how the people of a county can come up with the money to build and operate their own depolymerization facilities.

Can governments and oligarchs throughout the world be convinced that they must do this? If governments won't, what can we do? We can let our governments know that they can use our tax money to pay for this infrastructure. We may have to do street demonstrations. We could hold signs saying, TAX ME TO SAVE THE PLANET. We can vote for politicians who are willing to build such an infrastructure. If our city, county, state, national, or world governments will not take on this challenge, then the people must do so.

3. Worldwide Nuclear Meltdown

The extinction of the human race due to a massive nuclear meltdown is a real possibility. When the Earth's resources are no longer enough to provide for the population, the stresses on the economic and social systems of the world will become so untenable that humanity will have no recourse but to revert to the basic functions of providing elementary needs. At this point, humanity may not have the time and resources (such as electricity) to properly maintain the technological infrastructure that is used in nuclear reactors. If meltdowns occur on a massive scale, enough radiation may be emitted to kill off all life on Earth. The fail-safe way to deal with this possibility is to decommission these nuclear reactors. We need to begin the actual work of safely disassembling these facilities while we still have time. If we can start replacing enough nuclear-generated electricity with electricity supplied by wind and solar, then we will no longer need nuclear technology.

Who decides to decommission a nuclear reactor? This decision lies in the hands of domestic and international governments. Get ready for demonstrations. Vote for politicians who want to get rid of nuclear power. Talk to your friends. Convince everyone that solar and wind power can replace nuclear power.

4. The Need for an Infrastructure of Solar and Wind Power

The roof of every building owned by a government, business, or homeowner should be required to have a solar array. Every highway right-of-way that is suitable for solar panels or wind turbines should be used to generate electricity. Electricity from wind and solar is the energy of the future. As we plan for the future, electricity from wind turbines and solar panels will replace electricity from nuclear and fossil fuels.

What is required to build such an infrastructure? It will require the will of the people and lots of money and resources from everyone, including corporations and oligarchs. Can this be done on a piecemeal basis, where people voluntarily convert their own properties, or will it take a mandate and financial help from the government before people will agree to use their properties for the good of themselves and others? I'm afraid that people of all strata in every society will be hesitant to do anything unless they are forced to do so.

CHAPTER 4

Responsibility versus Freedom

SO HERE WE ARE WITH all of the solutions in hand. We know what needs to be done, from building hundreds of thousands of depolymerization facilities, to installing wind and solar arrays on every building, to dismantling all of our nuclear power plants, and to agreeing to limit ourselves to one child. Will we be willing to do all of these things? What would the consequences be if we do not? If we do not wage our fight against the worst-case scenario, I see a future where death occurs after a chaotic and horrendous time of unimaginable hardship and suffering for our families, our children, our friends, and our neighbors.

"But," you say, "there is no absolute proof that the end of civilization will happen as you describe it." To this

I say, "Why take the chance?" It will not hurt us to build all of these depolymerization facilities or an infrastructure of wind and solar arrays. If humanity can avert the current crisis, the world will still need the infrastructure that will have been built. The people of the future will need to regulate the amount of carbon in the atmosphere and will continue to generate electricity by wind and solar energy.

Do we really need to reduce the world's population? Think of it this way: when humans emerged on Earth, the world had an abundance of plants and animals. It was devoid of humans. Now, for the first time in the history of the planet, the human population has grown enough to fill the planet. Think of a glass of wine. You want to serve your guest, so you begin pouring wine into the glass. You won't have a problem as long as you stop pouring before the glass is full. If you are not paying attention and continue to pour and overfill the glass, you will have a mess on your hands. The moral of the story is that you have to stop filling the glass before it is too late.

Is Earth reaching its maximum population at this time? Has Earth already surpassed its maximum population? Are we ready for the mess that some scientists predict? We can avert the mess by reducing the human population to a sustainable size. Then the people of future times will have to continue to regulate the size of Earth's population so that it is always in sync with the available resources. The future of Earth will be a future of regulation and responsibility.

What a bummer! I don't like being regulated any more than you or anyone else. But regulation must be the way of future generations. The future brings with it a new paradigm. The planet has already been filled with people, and these people have depleted much of the planet's resources. Future generations will need to make decisions. How many people can Earth sustain? How do they make the best use of the resources they have? With a history of coming so close to extinction, the only choice they will have will be to accept strict regulation of population and resources. If we can get past today's crisis, the future will be just fine. A sustainable population with a sustainable supply of resources will live on a planet where the threat of extinction will never again be imminent.

But humans often seek the freedom to do as they choose. Whether in politics, religion, or everyday activities, many people assume that they have an innate right to exercise their liberties freely, without regard to the consequences. Now, suddenly, that will not be the way for people to progress. The way for people to progress will be for humanity to promote and accept governance that will establish sustainability. In the future, governance must be fair, and people will need to conform.

People will need to conform! People don't want to conform! People want to do as they please. People want to live in the moment. They don't want to be concerned about the future; they want their personal life experiences to stay the same, day after day, year after year.

The good news is that with a little bit of conformity, their lifestyles will be able to stay much the same as before. The necessary decisions will create challenges for everyone; however, everyone will need to do as instructed and work together as needed.

As we begin the struggle to remove carbon from the atmosphere and reduce our population, we will often have to be compliant. During these times, we will know that being good and helpful soldiers—rather than claiming unproductive freedoms—will carry the day. The specter of extinction hovers over all of our actions. We need to adopt attitudes that will enhance our fight rather than diminish it.

Having to act responsibly is a form of constraint. All people living on Earth will need to understand that they must stand ready to do the work of saving the planet, and they must do it without complaint. When one looks at the specter of total annihilation—the end of the human race—one fears the future and is awed by how much responsibility will be placed on each of us.

These are the responsibilities that everyone will need to accept: (1) to reduce the population on Earth by accepting stringent birth-control rules that will result in only one child per family; (2) to build an infrastructure of hundreds of thousands of pyrolysis/depolymerization facilities worldwide; (3) to use the roofs of all government buildings, all private business buildings, and all private residences for generating electricity by solar arrays and to use every highway right-of-way for solar and wind arrays;

and (4) to tear down all nuclear power plants because there will be no need for them in the future.

If we are going to save our civilization, then it will be up to our leaders in government and elsewhere to make these decisions. Everyone worldwide will have to decide as individuals to participate in turning these responsibilities into realities.

Who are these good folk among us who will take on the responsibility of saving civilization? I believe that action must be taken by the following worldwide groups: (a) scientists; (b) leaders from all religious orders throughout the world; (c) government leaders, from national levels to local levels; (d) leaders from every kind of business, whether it is for profit or nonprofit; (e) worldwide nongovernmental groups such as the United Nations, the World Bank, and the International Monetary Fund (IMF); (f) military forces everywhere in the world; and (g) the common folk. The *common folk* are all the rest of us. I will explain my thoughts on these various groups in the following chapters.

CHAPTER 5

Scientists

THE PROBLEMS WE FACE ARE huge, global problems, so the solutions must be equally global in size to match. That doesn't mean the solutions need to be mind-bogglingly complicated; the simpler the solutions are to comprehend and work with, the better. But the solutions must be accepted and administered everywhere in the world.

Who is to carry the banner and do the hard work to convince all the people of the world that they need to change their lives to become part of the solution? The answer is our leaders—but more especially, the scientists among us. And there is good reason why. The last report I heard was that nearly 100 percent of scientists know that global warming is created by humans.

Scientists are smart people, but they have a strange quirk: they are unwilling to say that anything is for certain unless it passes their stringent rules for scientific proof. All scientists, everywhere in the world, in every scientific position and from every scientific field, whether they are teachers or hands-on practitioners, need to rise and shout for all the world to hear: "We have enough proof. An empirical count of nearly 100 percent is enough proof. We need to be prudent and correct this anomaly called 'global warming' as quickly as we can. We don't have to wait to start. We can start today. We have solutions at hand, and we need to begin employing them."

Scientists need to shout on radio, on television, and in written media. Scientific journals would be a good place to start. Students in college and university classrooms must be kept apprised of our success in abating global warming and overpopulation. Students in high school and elementary school should be exposed to the topic to some degree; after all, it is their future we are talking about here.

The scientists will have to do the heavy lifting, but that does not let the politicians, business leaders, clergy, or anyone else off the hook. Everyone will need to know the solutions. Everyone will need to know that the solutions will work, and everyone will need to contribute to making the solutions work.

CHAPTER 6

A View from Wherever You Choose to Worship

RELIGION IS A VERY IMPORTANT part of this discussion. Religion will play a huge part in convincing everyone of the need to devote two or three generations of the forward march of humankind to healing the planet and getting humanity back on a sustainable footing. Solving the problems of overpopulation and global warming will require cooperation from all of us. At the present time, however, we don't have cooperation. Instead, we have competition and fundamentalism. All of the people of the world will somehow have to learn to cooperate rather than compete.

The problem is global. God is global. Cooperation must also be global. Globally, religion must play a huge role in creating cooperation among all people.

Perhaps I need to say a little about myself. I was born into Christianity. I was baptized into Christianity and was taught to believe in a triune Godhead—the Father, the Son, and the Holy Spirit. I know that other people have been born into and have learned to accept perspectives of God that appear to be very different from Christianity. I have come to accept, as my own belief, that God has revealed Himself to different peoples at different times in different ways. I believe that it makes no difference whether you are Christian, Jewish, Muslim, Hindu, Buddhist, Sikh, or one of many smaller groups. God is with you so long as you recognize that the manifestation of God in your religion is a spiritual being rather than an inanimate object or such concepts as greed, power, money, pride, and licentiousness. God has set forth laws and rules so that all of humanity can govern themselves. The people of all religions have received God's rules.

I can speak only of my Christian upbringing when I say that I try my hardest to adhere to commands given by God and pronounced by Jesus. Some of these are as follows: There is a place in heaven for all who believe. Do unto others as you would have them do unto you. Forgiveness of others is an act worthy of God and all humanity.

I bring to you my feelings about religion in order to reinforce the idea that if we are to end and reverse global warming, then all of the people everywhere in the world, regardless of their personal religious beliefs, must make personal commitments to accept all people as coworkers in the effort to end global warming. To save humanity from the specter of extinction, this commitment will have to last for two or more generations.

This mandate is of an especially high order because it involves people's beliefs. Everyone will have to welcome and even embrace the people of other religions. Even though the precepts and conditions of other religions are different from their own, people will have to hold their personal beliefs close to their chests as they attempt to communicate with others. The challenge will be for everyone to cooperate fully in taking on the work of saving the world and the people on it.

This will be difficult for those who subscribe to fundamentalist beliefs and also for those who have strong end-of-the-world beliefs, or what is known as "eschatology." Even as we acknowledge that some religions include a "rapturing" of souls into heaven as part of their belief systems, I submit to you that the extinction of humanity caused by overpopulation is not the Armageddon spoken of in scripture.

God gave us the planet and all that is on it. He also gave us the responsibility to be good stewards of what He gave. If we are unable to prevent extinction because we are too late, then we can know it was caused by human behaviors such as greed for power and money, a deliberate unwillingness to understand overpopulation, and an uncaring attitude toward things that affect people who appear to be different from us.

How should we consider the notion that the extinction of humanity might be brought on by human overpopulation? Did God not command Noah and his family to "go forth and multiply?" He did so in the Old Testament book of Genesis. But earlier in the book of Genesis, He gave His first instruction: He gave dominion (which implies stewardship) of His creation to humankind. As far as I am concerned, then, the first instruction, to be good stewards, takes precedence over the second, to go forth and multiply. I know of no scriptural reference that instructs humankind to overpopulate Earth for the purpose of causing our own extinction. Regardless, since I have no idea what God intends for the end of times, I will try to maintain Earth as the beautiful globe He made and has given to us.

Yes, we must all work together. We cannot let religious precepts stand in the way of reducing the amount of carbon in the atmosphere and of reducing the population; instead, we must require that our religious leaders become active in the quest to do these things. Religious leaders need to learn how population growth will affect Earth's resources and the probable consequences thereof. Religious leaders—especially those at the highest levels of their orders—need to engage with scientists and business leaders to determine a way forward that will result in reaching the goals of reducing the population, reducing atmospheric carbon, and dismantling nuclear reactors.

CHAPTER 7

What about the Oligarchs and Corporations?

As STATED BEFORE, THERE ARE scientists and other people who are strongly of the opinion that the human race will go extinct in the following steps. First, Earth will finally become so overpopulated that the economic structure will fail. Second, the social structure will fail. Finally, the technology structure will fail. When this fails, the nuclear power structure will fail, and nuclear power plants will melt down. The resulting radiation will kill all life on the planet and will leave a world that is devoid of people.

Because the scientists and other people who are certain that this scenario will play out are not a bunch of weirdos, we must consider their assertions in a thoughtful manner. At any rate, the rest of us are left with the following choices: we can disbelieve them and do nothing, or we can do something whether we believe them or not. What harm would be done if we ended global warming and made certain that the above scenario would not play out? To work at ending global warming would seem to be the prudent thing to do.

Now, what of the oligarchs and corporations? The oligarchs and corporations have acquired huge amounts of money; trillions of dollars have found their way to tax havens around the world. I don't know or understand how this money is being used. Is it being used to make more money for the oligarchs and corporations? Is it being used to do good deeds for others? I do understand that taxes are owed on this money; otherwise, the money would not be in a tax haven. Whether the taxes are owed to the United States or to some other country is not the point of this book; the point is that the money is not being put to use to solve the problems of overpopulation and global warming.

How should the money be used? Our business leaders need to think of the consequences of the scenarios stated above. Will our economic system fail and take the social structure and nuclear power plants with it? Do corporations and oligarchs have a responsibility to protect humankind?

Here is another scenario. What if all of the money on Earth was sitting in a vault when humanity became extinct? Night would still follow day, and the money would still be sitting in the vault. There wouldn't be anyone to open the vault to look at it or to spend it. How sad would that be?

It is reasonable to expect corporations and oligarchs to use that money to build the necessary wind, solar, and depolymerization infrastructure. Just think, if corporate leaders and oligarchs saved the planet now, then in the future, private, for-profit corporations like them could continue to operate as if nothing had happened. But if they do *not* use the existing money to save the planet, there won't be any corporate leaders or oligarchs to operate future private, for-profit corporations because oligarchs, leaders, and corporations will all be extinct. Oligarchs and corporate leaders can make contracts with governments, banks, and other entities to be repaid after they have solved the problem of global warming. Corporate leaders and oligarchs have to be prudent, as I've used the term above, and release their money immediately. If global warming is not solved, it won't make any difference what is written into a contract—because there won't be any people to read the contract.

Corporate leaders and oligarchs also need to understand that the unfettered, laissez-faire, free-enterprise approach to capitalism, in conjunction with a continually increasing population, has created a formula for an economy that can easily fail. In my mind, it would be better to have an economic structure that features a strictly controlled form of capitalism with a larger dependence on cooperatives and nonprofit corporations.

CHAPTER 8

What about the UN, the World Court, and the IMF?

GROUPS SUCH AS THE UNITED Nations, the World Court, and the IMF are political as well as economic entities. They have a lot of power, and they control a lot of the wealth that will be needed to solve the global warming dilemma. These groups will have to change the rules they currently work under so that they can supply the people who are working on resolving the global warming problem with the resources they need to do their job.

I'm not an economist, but it seems to me that in a capitalistic society, there is a natural flow of money from the bottom to the top. Each person spends a small amount of money, but it all flows into the hands of a few entrepreneurs at the top. If this money was redistributed back down to the masses as wages for working on projects that heal the planet, then the masses will spend it again, to the benefit of the entrepreneurs at the top. The entrepreneurs will not be hurt. The masses will not be hurt. The planet will be improved. But this hasn't happened. Please tell me why.

More has to be done. Banks and courts and political organizations will need to see that all of this money is found and put to work. Unfortunately, no one wants to give up his or her money without promises that the money will be returned. We will expect the wealthy to be cooperative as they hand over their money to be used to heal the planet. Of course, records can be kept, and the money can be repaid after global warming has been solved. Who will make and keep the records? Who will repay the money? And how will the money be repaid? My guess is that all of these questions will be answered in some kind of negotiations at the time that the money is needed. What is the most important thing to consider? If we are going to resolve the issues of overpopulation and global warming before it is too late, then we must begin soon.

The goal is to avert the worst-case scenario, which is the total extinction of life on Earth. For many people, the notion that all life on Earth could end is more than they want to think about. The problem is that we have no way of knowing that if we continue doing things the way we do them, then the worst-case scenario will not come to pass; we only know that if that should come to pass, then the rich and the poor will become equal. This is the time in the history of our planet that all humans, rich or poor, must be willing to work together. Nobody should be harmed, however, and the money should be repaid to the owners after the effort has been successful.

Greed and other sociopathic behaviors must be set aside if the war against global warming is to be won. Everyone must work to end global warming. This will take a long time, and we must get started soon. Organizations such as the United Nations, the World Court, and the IMF will need to be leaders in putting all of the wealth of the world into fighting global warming.

CHAPTER 9

What about the Military?

SOME MIGHT SAY THE MOST important function of the military is to make peace. Others might say it is to make war. Today, things are different. Please let me explain. All of the experience gained by military leaders since the beginning of time was gained when the population of the world was still growing and had not grown to the extent that it has today. Today, the prospect that runaway population growth will drive the human race to extinction is real. This changes everything about the role that military organizations must take in this fight against global warming.

What should a military leader think about when he or she is presented with the specter of extinction? How can people justify winning a war if their side will soon be dead anyway? The fight must be against global warming. I would ask our religious, scientific, political, and business leaders to instruct military leaders everywhere in the world to begin talking with their enemies. Tell them that it makes no sense to take the chance that they might die in a war when they confront the certainty that they will die from extinction. Eliminate the specter of extinction first. Wage peace until the world is safe.

Using depolymerization to remove carbon from the atmosphere will take some time—perhaps fifty or more years. This is how long we need to maintain peace. It will be up to governments (and, by extension, to military leaders and fighting forces everywhere in the world) to make peace a priority and to direct their efforts toward winning the war against global warming.

CHAPTER 10

What about the Common Folk?

THE *COMMON FOLK* ARE THE people who make up the majority of the population of any society. These are the people who do the work and get the job done. These are the people who are compliant enough to take orders and carry them out without making a fuss. These are also the people who will stand up for their rights and make a ruckus if they feel that joining a protest or demonstration will protect or improve the lives of their families and loved ones.

I believe that common folk are loving and concerned people who want to do the right thing. I think common folk will understand the implications of overpopulation when it is explained to them. The common folk will stand ready to do whatever is necessary to save the planet from nuclear annihilation.

They will understand that the population of the world must be reduced. They will allow themselves to accept reasonable rules for population control. And they will act on those rules because it will allow the population to be reduced in the most humane way possible without having to resort to draconian measures. The common folk will understand that their grandchildren and future generations will need time to build the planet back to the lush provider of all things that God has given to the human race.

They will do the right thing, but they will need to be told to do so by leaders they trust. People put the greatest trust in their religious leaders. Religious leaders have to be onboard so that everyone will be fine with doing everything they are asked to do. The common folk will support the efforts of religious leaders, scientists, businesspeople, and politicians as these leaders engineer the solutions to the planetary problems we face.

CHAPTER 11

An Action Plan and the People to Work It

As you know by now, this book is about controlling the population, ending global warming, and dismantling nuclear power plants. I wrote this book because it is plain as day to me that overpopulation will tax the planet beyond its ability to supply the needs of the people. When this happens, the economic system will break down. Soon after, the social structure will break down, and families will become stressed beyond reason.

What will happen when people begin looking after themselves and their families only? They will look after their friends and families before they look after their technology. There is a strong possibility that the technology system will break down. If we lose the technology structure, then all of the nuclear power plants around the world will be jeopardized and may fail. As I mentioned earlier, some scientists say that if all of the nuclear power plants in the world melted down, there would be enough radiation to bring about the end of all life on Earth. That is the worst-case scenario.

A better scenario would be if the nuclear reactors did *not* melt down. In that case, although the world would still descend into chaos, with widespread deaths and hardships, humanity would survive to pick up the pieces.

We have several choices: (1) we can accept that the worst-case scenario will happen and wage a total, full-blown war against that specter; (2) we can do only the minimum of dismantling the nuclear power plants so that we at least have that protection; or (3) we can do nothing and accept the consequences of our inaction. Against which scenario should we stage our fight? To be prudent, we must fight the worst-case scenario. This means we must reduce our population, reduce the amount of carbon in the atmosphere, and dismantle the nuclear power plants so that they will be safe if the technology system fails.

We have a lot of work to do.

Who will do this work? Who will be in charge? How can this be done worldwide? This is a herculean task. Over seven billion people must work together with the same purpose. How can one even think about the enormity of such an effort? Nevertheless, the effort must be made. People will have to work together and follow orders.

How will we start? What follows is a rendition of how I hope things will work out. Please note that I have written each step as if everybody is doing the right things and we are winning the war against global warming. This is not wishful thinking on my part. It is an effort to give you the psychological boost you may need to help you get involved in the solutions. Keep a positive attitude. Think in terms of winning this war against overpopulation and global warming. Then get out there and do everything you can.

OUR RESPONSIBILITIES
First, I'll describe a series of steps that people can take; I'll then describe a series of logistical steps that we can take to ensure that the "people" steps progress smoothly.

PEOPLE STEP 1
We will start with our religious leaders. The pope, the Dalai Lama, and the rest of the highest-positioned leaders from all of the world's religions will meet with the world's leading scientists. These religious leaders will come away from their meeting accepting that one male and one female should bear only one child, and they should do so by adhering to the new, strict birth-control rules set out in addendum 1. This situation would last for two or three generations. China has tried this, but in a different setting and with different rules. What

happened, and whether they were successful or not, are irrelevant. Going forward, what we will have to acknowledge is that it will be scary for everyone. Religious leaders will have to be there to say that limiting a family to one child is OK. People will comply if they see the specter of extinction and if they know that what they are expected to do is OK with God.

People Step 2
Religious leaders will, in conjunction with scientists and the United Nations, meet with the top leaders of the world's nations. The overriding specter of extinction will be with them at the meeting table. The nations' leaders and legislative assemblies will see and understand that extinction is imminent. They will make laws and rules to decommission nuclear power plants, to reduce the population, and to remove greenhouse gasses from the atmosphere.

People Step 3
Religious leaders, scientists, and politicians will meet with oligarchs and corporate leaders. Again, the overriding specter of extinction will be with them. At the meeting, business leaders will come to the decision that all available money will be used to build infrastructure for solar and wind generation, as well as depolymerization facilities. Contracts will state that the money will be paid back, even if not for fifty or more years.

Business leaders and oligarchs will realize that they have no choice but to ensure that the war against the worst-case scenario is fully funded. They will see the specter of extinction and realize that money won't do anybody any good after the human race is annihilated. They will agree that all of the resources of the planet will have to be put to use in a life-and-death struggle to prevent the extinction of humanity.

People Step 4

All homeowners throughout the world will have to allow food to be grown in their yards. This new way of life will have to go on for at least two generations. I believe that people will become used to this new lifestyle and that they will end up having better food and healthier diets than we do today.

People Step 5

Everyone, worldwide, will have to be of the same mindset. Making the decisions and doing the work to resolve the worst-case scenario seems daunting, but I am satisfied that it can be done. Doing anything on a worldwide scale is almost impossible, but this mission is of such grave importance that everyone will need to stay focused on his or her part of the solutions. For the common folk, our jobs will be to give birth to only one child for every two people, allow our yards to be used to grow food, and be the compliant labor force that will do the physical work of the change. The experts will take care of the details. This is as it has always been. We common folk will see the specter of human extinction staring us in our faces and will know that it is real. We will simply do our best to do our jobs.

THE ACTION PLAN
Here is where we will get into the details of how, exactly, we can prevent global catastrophe.

LOGISTICS STEP 1
At the earliest possible time, scientists, engineers, and corporations will begin building an infrastructure that consists of hundreds of thousands of depolymerization facilities. As we've discussed, this is the process that changes carbonaceous material into diesel fuel, other petroleum products, and electricity. These facilities will help to provide the energy that we now get from fossil-fuel sources.

LOGISTICS STEP 2
Farmland will be used to grow carbonaceous material such as switch grass, yellow sweet clover, and hemp. This is the feedstock that will be processed at mini-refineries to generate electricity and petroleum products. At this time the technology requires that pyrolysis/depolymerization

be done in individual batches. We will know that we are removing carbon from the atmosphere because a residue of carbon remains after each batch is processed. It will take many years to remove all of the carbon we have put into the atmosphere, but, with hundreds of thousands of facilities processing millions of batches of feedstock every day, it will be done.

Logistics Step 3
Automobile, farm machinery, and industrial manufacturing companies will put their factories to work building the pyrolysis infrastructure. They will get the necessary infrastructure up and running without delay.

Logistics Step 4
What will we do for food if we are using our farmland to grow the plants that will remove carbon from the atmosphere? We will be willing to use my plan (or some variation of it) as explained in addendum 3, "Front-Yard Gardening." We can sustainably grow our food in our own yards and on marginal land, thus providing abundant food for everyone.

Logistics Step 5
What about the economics of a smaller population? We are so used to a consumer society that we can easily

become afraid that our economy will fail if we do not have more and more shoppers. This thinking will change as we begin to have fewer shoppers and fewer people to make and supply the things we need. The experts will figure out how to make sure everyone can share in the available resources. I have always been of the opinion that a stable economy is a function of employment. The more jobs, the more stable the economy will be. As the population grows smaller, jobs will be available for everyone, provided that the amounts of goods and services stay in sync with the needs of the population.

It is not hard to understand that global warming is a consequence of overpopulation. Scientific research has already shown that too many people are adding too much carbon to our atmosphere. The problems of too many people and too much carbon must be addressed at the same time. The world's leaders in religion, science, government, and business must begin the process of change. I call on the community of scientists to teach religious and government leaders—in the most vocal means possible—the ramifications of the worst-case scenario. I call on religious leaders to pray for understanding and then to convince other leaders, especially business leaders, that making decisions that result in corrective action is something God expects from us. I call on government and business leaders to understand their roles in providing all of the necessary resources. Finally, I call on the common folk to do their part by helping to do the work, to understand the need to be compliant with new rules such as a one-child rule, and to be the watchers who will guard against any attempts by government and business to turn our fight against extinction into personal gain. After all, if we become extinct, what good is personal gain? In the beginning, God gave us this beautiful world. We cannot allow humankind to destroy the world and eradicate itself by the sinful acts of greed and neglect.

I might be but one voice crying out in the wilderness. Even so, this one voice implores you to recognize that there are three watchwords that apply to the effort we must make to save the planet and save humanity. These words are *prudence, prevention*, and *action*. We must be prudent and choose to fight our battle against the worst-case scenario. We must put all of our energy into preventing the final stages of overpopulation and unabated pollution. And we must begin as quickly as we can.

CHAPTER 12

A Plan for the Nations

IN CHAPTER ELEVEN I WROTE a most idealistic scenario: the richest oligarchs and the poorest common folk all pitching in to help solve global warming in any way they can; money being returned from tax havens; corporations gladly using their factories to produce global warming infrastructure; journeymen, apprentices and common labor all standing ready to do the work. We can only wish that human nature worked this way. The reality is: people don't do stuff unless they really believe or are pushed.

In chapter twelve I will lay out my action plan by which that needed push will be given. Action is what we need. So far, there has been more talk than action. We agree. We have a global problem. The fact that no person, no organization, no nation is in charge, stands out loud and clear. How can we solve global warming if no one is made the boss? Where can we turn to find someone to tell us what to do? My inclination is to turn to one of the most important people in the world.

In paragraph #175 of Pope Francis's encyclical on the environment the pope writes, with regard to the environmental crisis, "As Benedict XVI has affirmed in continuity with the social teaching of the Church: 'To manage the global economy; to revive economies hit by the crisis; to avoid any deterioration of the present crisis and the greater imbalance that would result; to bring about integral and timely disarmament, food security and peace; to guarantee the protection of the environment and to regulate migration: for all this, there is urgent need of a true world political authority, as my predecessor Blessed John XXIII indicated some years ago.' "

Pope Francis and his predecessors have it right, especially with regard to the environmental issues facing humanity today. Global warming does indeed include such topics as peace, food, immigration and economic stability. If I were the pope I would have added reducing the human population and ending nuclear energy production to the list. But regardless, the question is: "Would it be possible to create a _true world political authority_ as envisioned by the popes?" My answer is yes.

My suggestion is for the nations of the world to appoint a temporary world government, with authority limited to ending global warming. The goal of this temporary authority will be threefold: remove carbon dioxide from the atmosphere to a level of 280 parts per million; reduce the population in a humane, nonviolent, non-eugenic manner; and dismantle everything nuclear. These three goals when accomplished will provide a healthy and happy place for humanity to abide, for a very long time, if not forever.

The single authority called for by Pope Francis cannot be a single person, or a junta, or a cabal of military generals, or a dictator, because they will not respect the people. The single authority cannot be a religious sect because individual dogma can become so strong that it can become misdirecting. The single authority cannot be a single country, because it will put itself first; or a group of countries, because they will argue and delay amongst themselves. The single authority cannot be The United Nations, because proposals can be vetoed by one country in the Security Council. The single authority cannot be the International Monetary Fund or the World Bank or the World Court, or NATO because their work is authoritarian or military in scope. Our single authority must be trusted by the people. It must be given the authority to operate freely. The nations of the world must be willing to defer to it. Nations must be willing to use their strength, even their military might, to support and protect this single authority.

Please consider: the problem is global therefore the solutions must be global. The solution is a world government. I see a world government that is agreed to and put in place by the nations acting together. I see a world government that is controlled by the nations. I see a world government that has been given powers limited to ending global warming that will last only as long as the nations want it to last.

This means that the nations will have to make treaties among themselves. The treaties will say that the new temporary government is in charge of ending global warming; that the nations will not interfere; that the nations will provide, without question, all of the money, expertise, manpower, goodwill, police and military support that the new temporary government should ask for.

When the nations decide to act they will make certain that our new government will always have officers who are trusted by the people. There are two factions in the broad makeup of humanity that are trusted by the people. The people will trust a new temporary government that has religion and science/education at its core. People, around the world, trust their religious leaders and their teachers.

What will this world authority look like? It will have one ruling body to set out the work that needs to be done; a special committee to clear the work site of encumbrances; and a group charged with the responsibility to see that all of the work gets done.

Where will they find the people to run the world authority? My strict rules for choosing these people are: they must be trusted; they must want to end global warming; they must have an aptitude for leadership; they must not be seeking political nor monetary gain; they must not have other agendas. My choice of where to find people with these qualifications is in a remote nondescript place. The temporary government will have officers who will be chosen from the science and comparative religion/humanities departments of colleges and universities everywhere in the world.

Please note that leaders chosen from science and comparative religion departments are as qualified to administer the office of a special world government as people chosen from anywhere else. And, besides, universities have legal departments with the expertise to write treaties, constitutions and all those other niceties that will no doubt be needed. Colleges and universities are not weak forces in our societies. All of the colleges and universities, worldwide, acting together will become a strong platform, a strong force, on which to build a temporary world government. But, just like the nations, the colleges and universities will not be permitted to interfere with the temporary government. They will be expected to join in supporting and maintaining the new government.

I have given this new world authority three distinct parts. It will have a ruling body to set out the work that needs to be done; a special committee to clear the work site of encumbrances; and a group charged with the responsibility to see that all of the work gets done. Let me introduce you to the names that I have given these three bodies: the 'Alliance', the 'Committee of the Prominent' and the 'Association'.

Of Population and Pollution - A Global Warming Primer

The Alliance

This is the entity that the nations will anoint as being in charge. The Committee of the Prominent and the Association will serve at the pleasure of the Alliance. The Alliance is an alliance of science and comparative religion departments. At this early time in developing this process I have chosen to call it simply the Alliance. The nations can give it a formal name at the time of anointment.

After having authorized the Alliance the nations will appoint temporary officers whose job it will be to arrange a time and place for an election of permanent officers. Let me give you a feel for how this election process will work.

- A. When the nations authorize the Alliance they will agree that the only people who can hold offices and be in charge of the Alliance will be people chosen from the science and comparative religion/humanities departments of colleges and universities worldwide.
- B. Employees from these departments who choose to be candidates will meet at a place and time and elect officers from among themselves.
- C. There will be two elections at different venues: one for candidates from the science departments and one for candidates from the comparative religion/humanities departments.
- D. If it is decided that twelve officers should run the Alliance then six would come from science

departments and six from comparative religion/humanities departments.

E. Two college campuses will be made available, one for each election.

F. To start the election process two or more days should be set aside for meetings and conversation among the candidates; so that each candidate can get to know the others.

G. Each candidate will be given two votes to cast as they choose. They could cast one vote for themselves and one for another who they think might be an effective officer. Having the choice they could cast both votes for themselves or however else they may choose.

H. Electing 6 officers from a large number of candidates will take several ballots. The people setting up the election get to choose how many ballots will be needed and how the candidates will advance to subsequent ballots. The next three bullet points are three ways that I might suggest.

I. One way might be to eliminate candidates who received zero, one or two votes. Continue this process on subsequent ballots until only six candidates are left standing.

J. Fewer ballots will be needed if the exponential powers of the number of winners are used. For instance, if 6 officers are to be elected then the number of candidates on each ballot could be limited to not more than 1,296; 216; and 36; by choosing the candidates who have the highest vote totals on the previous ballot; until 6 candidates with the highest vote totals remain standing.

K. Still another way would be to use agreed upon multiples of 6. For instance, each ballot could be limited to not more than 1,200; 600; 300; 60; 30; 18; and 12 of the highest vote getters, until 6 winners are left standing.

L. Time will be set aside for more meetings and more conversation between ballots.

M. Those candidates who have been eliminated will be asked to stay to participate in voting for remaining eligible candidates.

N. If the task is to elect six officers, then the election will be over when six candidates are left standing after having received the greatest number of votes.

O. Who will be at this election? I expect that hundreds of colleges and universities will send their best, their most qualified staff, from the science and comparative religion departments, hoping they will become elected.

P. The election process will choose the best of the best.

THE COMMITTEE OF THE PROMINENT

After the Alliance has been staffed with permanent officers, its first task will be to appoint a very special committee. The entire notion that we will ever be able to end global warming will hinge on this committee. The people appointed to this committee are the most revered and most awarded people on the planet. Their job will be to convince everybody that global warming is real; that we have to stop using fossil fuel; that the planet is being overcrowded with people; that all of the wherewithal available, including money from tax havens and other accounts, must be made available; and that the people are kept apprised so that everyone has a positive understanding of events and efforts. This committee will be staffed with the most prominent people in the world and their emissaries.

The people serving on the Committee of the Prominent will need someone to stand in for them when they are not available. These are the emissaries. The emissaries will speak with the same authority as the peerless ones. The Alliance will determine the size and functions of this committee. This committee will be answerable only to the Alliance.

THE ASSOCIATION

Let me use a simple metaphor to help explain the Association. As we go about our daily lives we may decide to remodel, repair or otherwise fixup something that we own. First, we decide on the job. Second, we study the ramifications of the job and remove any contingencies or encumbrances. Third, we get to work and do the job.

Our new temporary world government will act in much the same way. The Alliance will decide what needs to be done; the Committee of the Prominent will remove the encumbrances; and the Association will get to work and do the job.

I see humanity as being divided into four subsets: science, religion, business and government. The Association is named because it is made up of the association of these four separate groups. The Association will develop and engineer solutions; organize business arrangements; negotiate contracts; assign work and workers; provide oversite to the workplaces; see the work to its completion and report back to the Alliance. People identified with these four groups have the abilities and characteristics needed to coordinate, arrange and complete all of the work that needs to be done to end global warming.

The Association will have a board of directors. The board will be made up of an equal number of members from each of the four divisions: science, religion, business and government. For instance, if the board is to have twelve members then three will be elected from each faction. Each faction will hold its own separate election following the same process as that used to elect the officers of the Alliance.

How will we know if a person who wants to become a board member of one of these groups is best suited to that particular group? The Alliance will set up criteria to direct candidates to the most suitable group. We can know and be assured that each faction of the board of directors will be made up of the best of the best in each of the factions.

After the elections, the board of directors will decide among themselves who should be the Chair, the Recorder and other needed positions. A constitution, bylaws, standards of operation and other documents would be written and approved by the Alliance. The Association will become a very large, bureaucratic entity; facilitating projects and work orders everywhere in the world. All of these projects will have been ordered by the Alliance.

Pope Francis asked for a true world political authority. Who in the world has the authority to create such a thing? The short answer is 'the nations'. The nations are of the people. The nations not only have the authority, they have the responsibility. National leaders must not shirk their responsibility toward their constituents with regard to global warming. The United Nations recognizes one hundred ninety five member nations. Wouldn't it be grand if one hundred ninety five nations could decide to act together in creating a temporary, limited, world government designed to end global warming?

The process I have laid out will work if the nations allow it to work. The nations will have to sign treaties to allow the Alliance to exist. Perhaps the United Nations will be able to help with that. Where will the Alliance and all of its activities be headquartered? This decision will be made by the nations. How many nations will we need to get the conversation started? Perhaps two or three will be enough.

Will the temporary government be easily corruptible? No. The election process will ensure that only the best of the best will serve. It would be possible, I suppose, that an official could be bribed or otherwise compromised. But a provision for special elections could control that.

We are short on time. We need to decide what we are going to do. Let me rephrase that. *We are short on time. The leaders of the nations need to decide what they are going to do.* I suggest that the leaders begin meeting together and create the Alliance and the Committee of the Prominent and the Association. If the nations give up a little power to a temporary, limited in scope, global government, they will have engineered mankind's best bet to save the planet from global warming. It can be done. The reality is that nations are permanent; the Alliance will be temporary. Setting all this up will be good work for lawyers from colleges and universities.

Everybody knows how important it is to end global warming, but, nothing will happen unless there is a boss to tell us what to do. What if the nations won't act or can't get together to act? If the nations won't do it then the people will have to do it themselves. The people can petition the presidents and boards of regents of colleges and universities. Science and comparative religion department heads could begin talking it up with their peers from other universities. They could be the petitioners.

Universities and colleges are not without power and prestige. If colleges and universities throughout the world sign on to having a single world authority, national officials will pay attention. When the national leaders see that support of the people is strong, they will be able to accept a temporary world authority. National leaders have a choice. They can use their own initiative or they can listen to the people. In either case a temporary government such as the Alliance must be approved.

Please consider the following excerpt from a speech made by Dr. Martin Luther King on August 28, 1963. This speech is often called his fierce urgency of now speech. "We are now faced with the fact, my friends, that tomorrow is today. We are confronted with the fierce urgency of now. In this unfolding conundrum of life and history, there is such a thing as being too late. Procrastination is still the thief of time. Life often leaves us standing bare, naked and dejected with a lost opportunity."

We really don't have much time left to deal with the problems facing humanity. I am concerned for my grandchildren, my greatgrandchildren and future unborn progeny. Why should they suffer the chaos and hardship that will lead to possible extinction because we have been unwilling to act? The specter of extinction of the human race looms and we must acknowledge. Be afraid. Be strong. Be willing to choose a method to combat global warming that includes: a temporary world government with leaders who can be trusted; a nonviolent, non-eugenic process to reduce the population; a plan to dismantle all things nuclear; an infrastructure of solar panels and wind turbines to end the use of fossil fuel; an infrastructure of mini refineries to remove the excess carbon from the atmosphere; and a gardening technique that will feed the population while crop land is being used to save the planet.

My hope is that this chapter will inspire the nations to create the Alliance for the purpose of solving global warming. The Alliance will be given the protection, the wherewithal and the authority it will need. These needs will include trillions of dollars as well as factories and suppliers. Providing the needed equipment, materials and manpower will be crucial. Please know that if every decision is filled with empathy and understanding; things will be done right; nobody will lose.

Earth and humanity are forever linked. Humanity began with only a few people and has now grown until it is no longer sustainable. Earth's resources are being depleted. This is all happening on our watch and we need to act. We must maximize energy output from renewable sources.

The sun is the source of the wind, the ocean's waves and direct solar radiation. Gravitation is the source of the tides. These are the renewables. Nuclear power is not renewable. It depends on a resource, uranium. Geothermal is not renewable. It depends on the heat from the center of the Earth. Hydroelectric power depends on gravity and is considered to be one of the renewables. But, wouldn't it be nice to allow our rivers to run free?

The future of humanity is dependent on the amount of energy we derive from the sun. It is the sun that provides life on Earth. Let us use the sun to the fullest extent.

If one wanted to employ a little dark, snarky humor one might say, "Earth does have a safe nuclear power plant. It's the sun. It is safe because it is ninety three million miles away." Another might say, "Yeah, but It'll run out of power in two billion years. Will humanity last that long?"

ADDENDUM 1

Reducing the Earth's Population

EVERYONE CAN AGREE THAT THE Earth's population has grown from small clans of families who existed on what they were able to gather for their sustenance, to larger groups who depended on their ability to produce their own food, to still larger groups who settled in cities, where their sustenance became dependent on others. Throughout this time of growth, the Earth was a sustainable planet. It was easy for the population to exist on the largess the planet had stored up since its creation.

I think everyone can agree that the Earth's population has now grown so large that we can no longer depend on the largess of the planet for our existence. We are growing ourselves out of our home, and we have no place to go to build a new one. As the population continues to increase, we will not be able to feed everyone, nor will we be able to house everyone.

How can the Earth sustain a population that continues to expand without limits? I suggest that it can't, and I suggest that everyone knows it can't. I suggest that scientists, businesspeople, lawyers, journalists, doctors, pastors, entertainers, restaurant owners, garage mechanics, hotel workers, politicians, and the people on the street know that the population cannot continue to grow without limit.

Look at nature. Whenever a population of animals gets too large, disease sets in and kills the population down to a size that is again sustainable. The human population will soon experience this very situation. We will experience increases in disease and pestilence. It will become harder to find enough food to feed everyone. Adequate housing will become increasingly problematic. The economic system we depend on will collapse, creating even heavier burdens, followed by a complete breakdown of our society. What happens afterward will be horrendous.

The weak and the lame will be the first to die off. The healthy and strong will then band together in gangs to wreak havoc on one another in an effort to gain the sustenance they need. It will be a time of "everyone for themselves," without regard to how many people suffer and die.

We will witness the consequences of this scenario as it unfolds. In fact, we are already witnessing them—global warming is one of the first consequences of overpopulation.

Not everyone agrees that global warming is actually happening. Of those who do, not all sense that global warming and overpopulation are related. So it appears that we have two jobs to do, beginning now and going into the future. We must undo overpopulation by reducing the number of people on Earth, and we must undo global warming by removing carbon dioxide from the atmosphere.

This book describes how humanity must go about the business of reducing the population of the Earth to a sustainable level. But first let us address how dire the situation is. Because we have allowed the planet to warm beyond the point of sustainability, the entire human race will likely become extinct. How might extinction occur? After our social and economic systems fail, causing a huge die-off of people, will we then be extinct? What is left of humanity will, we hope, be able to rebuild and start anew, but I don't think it will happen that way. A commonsense review of what I have read on the subject tells me it will happen in a way that no one is talking about. Total extinction will likely occur because of the simultaneous meltdowns of all of the nuclear power facilities in operation at that time.

This will happen because we live in a technological age, but the technology that we are all so in love with will fail. The people whom we rely on to operate all of our technology will be caught up in the same social and economic chaos as the rest of us. There is no reason to believe that the workers who operate our nuclear plants will be on duty at the plant or that the personnel who are expected to take care of "the cloud" are going to continue to do their jobs. Failure in one place will cause failures elsewhere, and in no time, the whole technology system will shut down. Our nuclear plants will be adversely affected.

Let me reiterate. Soon after our social and economic systems fail, the technology systems will fail, and immediately after that, our nuclear power plants will fail and melt down. It will be the radiation from the meltdowns, I predict, that will wipe out civilization. The specter of total annihilation and total extinction exists. If we are prudent, any action we take will be with that specter of total extinction in mind.

What can we do about this today? We can begin by reducing the population in order to take pressure off of the planet's finite resources, we can begin by removing carbon from the atmosphere, and we can begin by decommissioning our nuclear power plants. As far as I know, it takes a long time, perhaps up to fifty years, to shut down an active nuclear plant and make it totally safe. Humanity may not have fifty years. We need to begin these corrective measures immediately.

I submit that it should be our goal to decrease the population. If we are to avoid any of the above scenarios, we have no alternative but to reduce the population. I further submit that we can, with the help and understanding of all people worldwide, reduce the population in a humane way. The humane way is to allow each couple, mother and father, the progenitors, to give birth to only one child of their own descent and raise that child to adulthood. This requirement will not be simple, but it will be achievable. On the positive side, the parents and child will make a family that can expect to be happy and loving. Compare this to the uncertainty, fear, and distress of a collapsing economic and social infrastructure.

What rules might govern the transition from a society where everyone has the freedom to have as many children as they want to a society where only one child is allowed per couple? I have my own ideas about how this can happen. Read on.

It would seem that the simplest way would be to "tie the tubes" of the mother and father after they have given birth to their first child. But it is not that easy, because there are vast differences between men and women. Performing a vasectomy on a man is fairly simple; it can be done in a clinic, and the man walks out with only a small incision that heals quickly. It is not so simple for a woman. A tubal ligation is a major surgery. There is no way to expect to perform major surgery successfully on one-half of the world's population.

We will therefore have to be thoughtful and perhaps even ingenious in our rule making. We must remember that the specter of extinction looms over us as we consider humanity's options. The following are the rules I think we should have for the next one to three generations.

Because men can father children even when they are seniors and because a vasectomy is a simple surgery, every man will get "snipped" if he has already fathered a child or when he fathers a child. There will be no escape for the men. But then again, they will be done and ready to move on after one simple procedure.

It will be much harder for the women. Women will not have to undergo major surgery. After they have given birth, they will be required to sign a document certifying that they have had a child, that they will not have another, and that, should they become pregnant with another child, they will carry that child to term and give it up for adoption. There will be many cases where childless couples will be available as adoptive families.

We know and understand that some things never work out as expected. Some extraneous circumstances will always have to be taken into account. What if a married couple should divorce? Unless a court orders otherwise, the child would go to the father. This is because the mother will be able to remarry. If the man whom she remarries is divorced or a widower, then in either case he will have the child from his previous marriage. The woman would continue to be blessed with one child. What if the woman divorces and then remarries a man who has not been sterilized? She would be able to become pregnant and be blessed with one child.

There will be no abortions for the sole purpose of population control. Women will still be able to choose among many legitimate reasons to undergo a lawful abortion in places where the procedure is allowed, but not for population control only. Please allow me to reiterate: if a woman who becomes pregnant has already given birth, she will be required to give up the second child for adoption.

The rules stated above will need to be promulgated into laws and regulations for countries around the world.

These rules are not onerous, but they are different from the way societies have behaved since the beginning of time. People will accept these rules when they hear their religious leaders proclaim that it is the right thing to do. And, of course, we always have to remember that the specter of extinction is looming over us as we decide to do the right thing.

I don't intend to lay out every single nuance and consequence of how this program might work. I do expect that if leaders in religion and government consider this kind of program, a vigorous debate will ensue. Is vasectomy the best way to prevent childbirth? When should a child remain with the mother? When should a child remain with the father? This debate will be about the rights of men, women, and children. Laws will be promulgated, and courts will become involved. The rules for a program to reduce the population by at least one half will come out of this debate.

I do intend to say the thing that is most important: Humankind is overpopulating the world. We need to find a way to reduce the population. We must not debate for very long. We must figure out the best way to do this as soon as we can.

There will undoubtedly be naysayers. People will say we are off our rockers, that global warming has not yet been proven, that the world cannot afford the cost of removing carbon from the atmosphere, that having fewer people to buy goods will be harmful to the global economy, that extinction fits into God's plan for the end of the world, that it's too late anyway, and more. But I think that with careful planning and the innate willingness of all people worldwide to support their friends and neighbors, humanity will be able to afford to build a large enough infrastructure to keep jobs for everyone and to maintain a stable economy, even as the population is decreasing.

Global warming is not in God's plan. I learned in Sunday school that God gave everyone a free will that should be used for good, but it can be wrongly used to harm oneself and others. I believe that global warming has been brought about by humanity's free will to choose to be greedy, stupid, and self-centered.

In order to save the planet, we need to remove carbon from the atmosphere and reduce the human population. Carbon can be removed by deploying a sufficient number of low-tech, carbon-negative depolymerization facilities. The population can be reduced in a humane manner by allowing each couple to have only one child. The population can be reduced (in only one or two generations) to a size that the planet can sustain.

This is the way to save the planet. Let us do it in a positive, safe, and humanitarian way. I don't want to put my grandchildren and great-grandchildren through the chaos of an economic and societal breakdown followed by a mass extinction. I love them too much.

But we do have to get the word out to the population, including our beloved families, that for one or more generations there will be only one child per family and that other lifestyle changes will also have to occur. The bottom line is this: Everyone will have to be willing to work together. All of the things that need to happen can be made to happen if everyone is working to make them happen.

Every person around the world will have to understand that there will be nothing more important than working to solve global warming and that they have the ability to do the work. For instance, operating the machines and managing the facilities where we will change carbonaceous matter into the energy that will replace fossil fuel will not be difficult or onerous. Neither will growing and harvesting the feedstock material be difficult. The work will be apprentice- and journeyman-level work. There will be no need for a high level of technology. But there is a need to begin the work of saving humanity. This work can and should be started as quickly as possible.

What can we expect the future to be like? Once the goal of healing the planet has been reached, good things will happen. Families will be allowed to have two or more children, depending on the circumstances of that time. The people of the world will never again have to worry about global warming, because they will be able to regulate the amount of carbon dioxide in the atmosphere; they will even have an infrastructure in place to do so.

In the future, Earth will provide for all its inhabitants. It will be a happy world, free of worry and undue stress. But it will never come to pass if the people of today cannot be won over to accept the drastic changes they will have to make.

ADDENDUM 2

Science and Science Deniers

IN THIS ADDENDUM I WILL discuss some simple science that everyone needs to understand. I will also discuss those people who, for whatever reason, choose to deny simple scientific evidence.

SCIENCE

The consensus among scientists is that there was a time during the formation of the planet Earth when there was a huge abundance of carbon dioxide in Earth's atmosphere. Why is it important to know this? Because something happened. There isn't a huge abundance of carbon dioxide in Earth's atmosphere anymore. We need to know what happened to it.

We were taught in elementary school that when plants grow, they take in carbon dioxide and give off oxygen. This is the process of photosynthesis. The leaves of a plant take in carbon dioxide, the roots of a plant take in water, and the energy from the sun combines the carbon dioxide and water in the following formula: $6CO_2 + 12H_2O \rightarrow C_6H_{12}O_6 + 6O_2 + 6H_2O$. This formula tells us that when the energy from the sun combines six carbon dioxide molecules with twelve water molecules, the plant gains one simple sugar and has six oxygen and six water molecules left over. All of the carbon that entered into the plant as carbon dioxide has been used up to make the sugar molecule. The plant keeps the carbon and gives off oxygen and hydrogen. We see that the carbon that was in the air around us as carbon dioxide has now become an integral part of the plant. That is what happened to the before-mentioned abundance of carbon dioxide. Nature stored it in plants.

The magic of nature uses that one sugar, other nutrients, and a genetic code to create the vast diversity we enjoy in the world of plants. Some plants are beautiful to the eye and nose. Some are ugly and smelly. Some taste good, and some taste bad. Plants are complicated things. I am not going to explain the complexity of plant matter except to say that it is made up of tiny bits called "polymers." A word about polymers: even though polymers are "tiny bits," some tiny bits are more complicated than other tiny bits. The more complicated polymers are called long-chain polymers; the less complicated polymers are called short-chain polymers. Plants are usually long-chain polymers, while petroleum products such as gasoline, diesel fuel, oil, and natural gas are short-chain polymers.

Why is it important to have a discussion of long-chain versus short-chain polymers? We need to understand that nature, using the natural forces of heat, pressure, and geologic upheavals over millions of years, changed the plants that grew and died millions of years ago by breaking down the long-chain polymers of the plants into the short-chain polymers of petroleum products such as coal, petroleum oil, and natural gas. This is where the fossil fuel we burn as energy today originates.

The next step in this little science lesson is to discuss what happens when we burn the fossil fuel to heat our homes, propel our cars, generate our electricity, and more. This discussion is quick and simple. When petroleum products are burned, the carbon that went from the atmosphere into the plant is now recombined with oxygen and becomes carbon dioxide. The carbon dioxide cycle has been completed: from carbon dioxide in the atmosphere to carbon in the plants to carbon dioxide in the atmosphere.

Who would care—except that carbon dioxide in the atmosphere is a greenhouse gas that is causing a problem. Might there be a solution to this problem? Wouldn't it be nice if we could harness photosynthesis by changing the long-chain polymers *of newly grown plants* into the short-chain polymers of petroleum products such as gasoline and diesel fuel? If we could only mimic nature, we wouldn't have to use fossil fuel as our source of petroleum energy. If we intend to reduce the amount of carbon dioxide in the atmosphere, we will need a method of breaking down long-chain polymers into short-chain polymers without having to wait millions of years.

"Oh," you wonder, "we already have a method, only it's not being used."

Well, I'll reply, "That's a damn shame!"

What is this method that will change plants into petroleum? It goes by several different names: hydrous pyrolysis, anhydrous pyrolysis, hydrous depolymerization, anhydrous depolymerization, thermal depolymerization, thermal conversion, thermal distillation, destructive distillation, TDP, and more. These all refer to the same process. They use heat and pressure to break long-chain polymers into short-chain polymers. *Hydrous* means "with water," and *anhydrous* means "without water." The process is explained in an essay I wrote about ten years ago entitled "Depolymerization Explained," included as essay 4 in this book.

Just as a side note, of all the above names, I like the name "thermal depolymerization" the best. The word "depolymerization" says exactly what happens when long chains of polymers are broken down. And the word rolls off the tongue in a pleasant sort of way, which can't be said for "anhydrous pyrolysis."

You may have noticed that when I have written about the kinds of plants that can be grown to provide feedstock (often called biomass) to be used in the depolymerization process, my choices have been hemp, switch grass, and yellow sweet clover. Let me explain why these three are my choices.

Industrial hemp is probably the best choice. It is a fast-growing, eight- to nine-foot-tall, oil-bearing, densely voluminous plant that has an abundance of biomass. A new crop needs to be planted every year. It will grow in poor or marginal soil and is drought resistant. It needs about half the amount of water that wheat needs.

Switch grass is a tall, thickly growing grass that creates a huge amount of biomass. A single planting will produce for up to ten years before the land needs to be cultivated and reseeded.

Yellow sweet clover is a plant that I remember from my youth, growing up on the prairie. It grows up to six feet tall and creates a huge amount of biomass, especially when thickly planted. It can be cut for biomass at the end of the first year and will grow back on its own the following year. The plant needs two years to produce seeds. One important thing is that this clover is a legume and therefore fixes nitrogen into the soil. When plowed into the soil, it makes a very good soil enhancer.

There may be other plants that are better producers of biomass than these. I am not an expert on any of these matters, so the experts may come up with better ideas. It just seems to me that these three, grown and harvested in some kind of rotation, will provide the necessary biomass and, at the same time, improve the soil.

Science Deniers

There are those among us who are global-warming deniers. Actually, they are best called "science deniers." This is really sad because this whole business of global warming (and the way we can resolve it) has to do with elementary-level biology and Earth science. Should anyone be excused for not understanding? Perhaps they were daydreaming about recess time. Global-warming deniers have no excuse for not understanding what is going on as far as the science is concerned. Global-warming deniers are caught up in wrong-minded agendas that can be detrimental to the real agenda of resolving the global-warming crisis.

Who are some of these deniers? There are scientists, engineers, economists, and others who say that we cannot remove carbon because the cost is too high. There are corporations that claim to have the ability to solve global warming by manufacturing and selling products that are designed to put less carbon in the atmosphere; they are deniers because they refuse to understand that we must not only reduce the amount of carbon dioxide going into the atmosphere, but we must also start *removing* it. There are politicians and others who champion schemes such as carbon taxes, carbon credits, and what is called "cap and trade"; they are deniers because they are thinking only of reducing the amount of carbon dioxide that is going into the atmosphere. I think all of these folks have been sold a bill of goods. For some reason, they seem to be afraid to look deep enough into the problem of carbon-dioxide emissions.

It is easy to find deniers who will say that depolymerization is not sustainable. They often base their comments on whether the facility is making enough money or whether it emits foul smells. Keep in mind that if you consider the specter of human extinction and act accordingly, you cannot think that way. How important is it that profits are being made or that odors are in the air when compared to the total extinction of humanity?

Respected scientists from respected universities are finally voicing their concurrence that a mass extinction of plants and animals is underway. They are calling this the sixth extinction. Scientists have identified five previous extinctions, including the extinction of the dinosaurs. Some of today's scientists are voicing their concern that the human population may be part of this sixth extinction. The scientific community as a whole must begin talking about the pros and cons of a sixth extinction as it relates to humanity.

People who read this book will find all kinds of things to say to deny my reasoning and approach to this vital matter of overpopulation. The deniers say that the world has always survived. They say that the population has endured massive blows—such as the bubonic plague, wars, and famine—and has always rebounded. They are not taking modern-day medicine and other advances into account.

Consider what could have been the worldwide consequences of the recent Ebola crisis if we had not had all the medical advances at our disposal. Nothing happened, and that's the point. Modern scientific, technological, and medical advances are allowing the population to increase. But even with all of these medical advances at our disposal, the specter of human extinction may indeed have the upper hand if we do not take a stand against the worst-case scenario.

I have no more proof than anyone else on my position on overpopulation. I am speaking strictly about the notion that there will be hell to pay when the number of people reaches critical mass in relation to the total resources available. It doesn't have to be like this. With a little planning and constraint today, we can make the lives of our progeny much better tomorrow.

ADDENDUM 3

Front-Yard Gardening

Rooftop gardening has always been an interesting topic. Backyard gardening is usually what a homeowner has in mind when considering whether or not to have a garden, and I suppose container gardening can be done anywhere. Why choose front-yard gardening? Does the concept of front-yard gardening have anything to do with growing food to eat, or is it only for growing flowers and bushes? In the near future, front-yard gardening will be all about growing food to eat—but with a twist. I believe that front-yard gardening will be crucial to growing enough food to feed billions of people. But first, let us investigate this twist.

In the future, front-yard gardening will also be done in the backyard. But that is not the twist. The twist is that the homeowner will not do the gardening. The gardening will be done, more or less, by remote control. I want to take you to the future to see my vision of how humanity will grow food to feed humanity. Then, afterward, I will tell you why we must do this. The story begins as I am walking along a street in a neighborhood, sometime in the future.

As I walk down the street, I see houses and yards on both sides, and I notice that rows of vegetables are growing in the lawns. Upon closer inspection, I see that the vegetables are growing in containers made of plastic or aluminum. The containers, which are about eight inches wide, eight inches deep, and three to four feet long, are laid out end-to-end in trenches across the breadth of the lawn. The containers are filled with a growing medium that is rich in the nutrition the plants need.

I speak with a homeowner who is watering the containers of plants. I ask who dug out the trenches and put the containers in the ground. He says that the local vegetable growers' association did that. I then ask if he had to know all about gardening, insects, and plant diseases. He says he didn't have to, because the association did all that. I then ask if he had to harvest the vegetables. He says he does not have to harvest the garden, because the association does that. He reiterates that all he has to do is water the plants and that his agreement with the association allows him to pick some of the vegetables for his own use.

I decide to find out where this mysterious vegetable growers' association is located, to pay them a visit. The homeowner tells me that there is a neighborhood station nearby; I learn that the association has lots of neighborhood locations throughout the city.

When I walk through the gate, I see a greenhouse, a large pile of compost, and a covered work area. I ask a worker what he does there. He says he has to fill containers with a growing medium and then plant the containers with the seeds, or starters, as scheduled by the master gardeners. He delivers the newly planted containers to yard gardens that have containers ready for harvesting. He also picks up those containers and brings them back to the work area to be harvested.

He explains that different vegetables are ready for harvesting at different times during the growing months and that the master gardeners maintain a rotation of different vegetables at different locations to control pests and diseases. In addition, he says that all of the plant residue and growing medium left in a container after the vegetables have been harvested are put into the compost pile for the purpose of controlling pests and diseases. I am pleased to learn that all of the old plant material will be recycled to make new growth medium for another batch of containers. Finally, he says that the harvested vegetables are delivered either to stores and farmers' markets for sale or to small, local companies for processing.

Back to the present: my little story is written as if it were in the future. But when is it too early to start planning for the future? I submit to you that the future is today. I appeal to people to consider all ideas for growing food locally. But my plan gives structure to the process and maximizes the potential nutritional values that can be gained with local efforts.

Why will we need to use our lawns to grow food? That's what fields are for. Here is the reason: our fields will have to be used to fight the war against global warming.

To reduce global warming, we need to reduce the amount of carbon dioxide in the atmosphere. This can be done by growing plant life to extract the carbon dioxide from the atmosphere and then by using a depolymerization process to break down the plant material into petroleum products and electricity. The rub is that we will need to use all of the farmland that we are now growing our food on for the purpose of growing the plant material that will best extract carbon dioxide from the atmosphere. The only places we will have left to grow our food will be our lawns and other marginal land.

Being able to grow nutritious food while our farmland is being used in the fight to end and reverse global warming is key to our effort. I use the term "vegetable growers' association" because I think a nonprofit cooperative may be the best business model to manage this kind of program.

Let me revisit the things we must do to end and reverse global warming. First, we must reduce our population worldwide. Second, we must complete the necessary research and development and then build and put into use hundreds of thousands of hydrous and anhydrous pyrolysis/depolymerization facilities. Third, we must use our farmland to grow plants that are heavy in foliage, such as the aforementioned yellow sweet clover, hemp, and switch grass. These plants will provide the raw materials for our depolymerization facilities. Fourth, we must be ready and willing, on a worldwide scale, to use both front yards and backyards, as well as any available marginal land, to raise the food required to feed the population.

ADDENDUM 4

Pope Francis's Encyclical on the Environment

Pope Francis released his encyclical on the environment on June 18, 2015, at the Vatican. I had just completed writing the first draft of this little book and had sent it to the publisher for its first edit a month earlier. I had included a section in which I implored the pope and the Dalai Lama, as two highly notable religious figures, to gather other worldwide religious leaders to begin pressuring noted scientists to start the work that needs to be done to save the planet. I had just gotten the manuscript back from the editor and was busily working on revisions when Pope Francis issued his encyclical. It is interesting to note that since the release of the encyclical, the Dalai Lama has said that because the problem is global, religious leaders everywhere must begin speaking out for change.

Because I mentioned the pope in my first writing, I think I should share my thoughts on his remarkable encyclical. And, because the Dalai Lama has spoken out about involving all religious leaders worldwide, his voice should also be heard loud and clear.

How fun would it be to be able to record an actual conversation between the pope and the Dalai Lama? I will never get that opportunity. But I do want to share my thoughts by way of a conversation. So I will have to be content to record a fictitious conversation. My characters are Reverend P and Reverend D, two Christian preachers from small, local congregations. Please note that "reverend" is a gender-neutral term. This conversation could be between two men, two women, or a man and a woman.

Rev. P: Good morning. I'm glad we were able to get together this morning.
Rev. D: Yes. I'm happy we bumped into each other the other day so that we could set up this meeting.
Rev. P: We agreed that we wanted to discuss environmental issues and what we should be doing in our churches.
Rev. D: And don't forget the churches on the national and world scales. What should they be doing?
Rev. P: OK, then. Who goes first?
Rev. D: Why don't we start with a little prayer, just to focus our thoughts?
Rev. P: Go for it.
Rev. D: Dear God, please send Your Holy Spirit to watch over the thoughts and words of the conversation we are having. Guide us so that we are in tune with the action You want the world to take. Keep

us focused on Your creation and the things You want us to do to help protect it from harm. And, Lord, we want to thank You for this opportunity to meet. In Jesus's name we pray. Amen.

Rev. P: So we've come together to talk about environmental issues—more specifically, the pope's encyclical on the environment.

Rev. D: Yes. But the pope included so many issues that I don't know where we want to start. I didn't bring any notes. Did you?

Rev. P: No. So let's start anywhere. How about corporations?

Rev. D: OK. The pope came out forcefully against corporate greed. Corporate leaders seem to be putting themselves first and the people last.

Rev. P: Yes. Corporate leaders are hoarding money that should be available here and now for the problems of today.

Rev. D: The pope says that we all need one another and that all of humanity should be willing to help one another. Big, rich, multinational corporations are everywhere. But the rich countries are not helping the poor countries, and the people suffer.

Rev. P: The pope is right, of course. Greed for power and money is a strong force. It seems that money begets money. The rich and powerful become richer and more powerful, and they leave the poor folk behind.

Rev. D: I agree with the pope, too. Apart from personally owned businesses, we would be just as well off if we used cooperatives and nonprofit corporations more, and free enterprise–style corporations less. Our manufactured goods and other goods and services would be just as good as they are now.

Rev. P: And I agree with you. The difference between cooperatives and corporations is not about goods and services. It's about how the profits are shared.

Rev. D: Yeah, I would go for a well-regulated and vigorously supervised system of doing business. It would also be more in line with what Jesus taught.

Rev. P: And what we should be teaching to our congregations. I try to teach that to be filled with greed for power and money is bad just as often as I teach that forgiveness and grace are good. I can only hope it's enough.

Rev. D: The pope also wrote about overpopulation, poverty, hunger, filth, litter, climate change, biodiversity, urban planning, fossil fuels, and more.

Rev. P: Yes, he did. He was certainly thorough. Each of those issues can be discussed independently. Do you want to pick one for discussion right now?

Rev. D: Sure, but I disagree that these issues are all independent. In my mind, everything is related to overpopulation. If one can reduce the population, one can mitigate the problems, including global warming.

Rev. P: I don't think Pope Francis will agree. He wrote about the problems of overpopulation, especially among the poor, but he writes that the population will be increasing into the next century.

Rev. D: I think he is concerned that people may get the idea that the only way to reduce the population is by allowing abortions or by using artificial birth control measures. I'm against abortions, and I think you are, too. As a preacher, I don't want to advise women to have abortions, but I will love them, forgive them, and counsel them if they do. Regular birth control devices? I'm OK with those. I think that if the truth were known, most Christians, including Catholics, use birth control when they need to.

Rev. P: I, too, hate abortions, and I pray that the advice being given to women would be to allow the fetus to grow, except when rape, incest, or saving the life of the mother is part of the picture.

Rev. D: Getting back to the pope and overpopulation, There is a kind of birth control that I wish the pope and the Catholic Church could adopt. Have you heard of a little book called *Of Population and Pollution*?

Rev. P: No.

Rev. D: OK. The author writes about overpopulation and presents solutions to overpopulation and global warming.

Rev. P: Interesting. I may need to get a copy.

Rev. D: His idea for reducing the population is to require that every male, worldwide, have a vasectomy after he has fathered one child. Doing so would limit each family to one child. Then, after two or three generations, families could be limited to two children.

Rev. P: Say what?

Rev. D: Yeah. He says that the planet is filling up with people and that future generations will have to use sustainable population practices.

Rev. P: Weird. I guess I could get used to it. But what does that have to do with the pope?

Rev. D: Hey. We both know how strong the Catholic Church is on birth control. Yes, and I don't think there is much you or I can do about it. The Catholic Church forbids vasectomies. I don't think you or I could convince the Catholic Church differently.

Rev. P: Yeah. Changing the Catholic Church? That's unlikely. But let me take a quick look on my phone. Oh—here it is, the Catholic Bible. Genesis, chapter 9, verse 1 says, "And God blessed Noah and his sons and said to them, 'Be fruitful and multiply and fill the earth.'"

Rev. D: OK? Where are you going with this?

Rev. P: Here is the King James Version. It says, "And God blessed Noah and his sons and said to them, 'Be fruitful and multiply and replenish the earth.'"

>See, they are different. The Catholic Bible says "fill," and King James says "replenish."

Rev. D: Yeah. I see that. There were fewer people on Earth in Noah's time than there are now. The population has been replenished and more. Now it's full.

Rev. P: Such irony. If this passage has been fulfilled, then Pope Francis can use the Catholic Bible to justify vasectomies as a way to control the world's overpopulation crisis.

Rev. D: Perhaps so, but I don't think men are going to want to get vasectomies. They'll think that doing so will destroy their masculinity. They'll be too proud. Men are vain when it comes to their manhood. But then again, pride is a sin. When given a chance to think on this, the pope and other religious leaders might accept that not being willing to have a vasectomy is against God's plan and that regulating the number of births to achieve a sustainable population is in God's plan.

Rev. P: I like this line of reasoning. I can't think of anything from scripture that would require humanity to have so many children that all of the children would die of hunger.

Rev. D: And if governments mandated vasectomies for all men and the churches OK'd the idea, then I think the people would go along with it.

Rev. P: What other things are in this book?

Rev. D: The writer argues that overpopulation will become such a problem that society will fall apart. He says that we are so dependent on electronics that if our technology system falls apart, we may jeopardize our nuclear power plants. He says if that happens, then, as the power plants melt down, the release of radioactive material may wipe the human race from the face of the Earth.

Rev. P: Do you think there is even a possibility of that happening?

Rev. D: I don't have any idea. How could I know? I'm not a scientist or an engineer. Scientists need to weigh in and tell us the facts about it. Then we could start figuring out what to do.

Rev. P: The pope didn't say much about nuclear power in his encyclical.

Rev. D: I know. Maybe he should have said more. I worry about nuclear waste and meltdowns. It happened in Japan.

Rev. P: I felt so sorry for the people in Japan. I agree. The pope should have said more. Scientists ought to be working on getting the most out of more benign methods of generating electricity, such as using solar panels and windmills.

Rev. D: I notice that you just called solar panels "benign." Does that mean that nuclear power is malignant?

Rev. P: Absolutely. And we should treat it as such.

Rev. D: A thought just came to me. In his encyclical, the pope talks about how the scientific community and the religious community need to begin to work together. I would suggest that the pope and the Dalai Lama invite ten stouthearted, renowned scientists to meet with them and other religious leaders. The scientists would be told beforehand that they would be asked the following question: "The extinction of the human race is so huge an event that it cannot be minimized; therefore, can you tell me with one hundred percent certainty that the human race will not become extinct from overpopulation coupled with the meltdown of nuclear power plants? If your answer is yes, then you must provide solid scientific evidence, as well as solid human and political reasons, to explain why you are not concerned and how you know there will be no problem."

Rev. P: Hey, that is such a great idea. If the scientists say yes and can report, with no doubt, that there is no problem, then life goes on and everything stays the same. If they say no, on the other hand, then they will realize that they can no longer put off the start of an all-out effort to save the planet from overpopulation and global warming. They will have to go to work.

Rev. D: And if the scientists say nothing, they are really saying yes—but without the 100 percent certainty requested. They will be disingenuous. They will be the worst kind of global-warming deniers.

Rev. P: Worse yet, any scientists who have been asked but don't come to the meeting, knowing that the question will be asked, will have to live with themselves in a shroud of guilt. Even though I will not know these scientists by name, I will forgive them and ask God to forgive them. I will pray that the Holy Spirit will be with them and that they find a way to put all of their energy into working to heal our planet.

Rev. D: I pray that everyone understands that when the stakes are so high, there can be no shirking of duty. The scientists need to show up on this one.

Rev. P: The pope did say one more thing of interest. He indicated that he was worried that nothing will be done because no one will be in charge. He alluded to the world having a "true world political authority" to direct everything that will need to be done.

Rev. D: He said that? That would be against everything that I believe in.

Rev. P: Me, too. Having only one person in charge of the world? How much greed for power can visit one person?

Rev. D: I don't think a single world leader is necessary. It will be up to the religious leaders, the scientists, and the people to see that everything is OK. They are the ones who will have to keep the politicians and business leaders in line.

Rev. P: I don't think the pope has to worry. You know, we are not dumb. I predict that religious leaders, from the highest of each order to the preacher at the smallest church, and scientists, from the Nobel Prize winner to the elementary schoolteacher, can and will work together. Add the rest of the people to that effort, and you will have a formidable force. The religious and scientific leaders will be hugely respected and dominant. The common folk will be a huge consumer and voting bloc. Together, they will make sure the politicians and business leaders do the right things.

Rev. D: I agree. This job is for the people to do.

Rev. P: So that means that we also have a job. When should we talk about what we can do?

Rev. D: How about next week...maybe Thursday at three, right here?

Rev. P: OK. Let's finish up with a short prayer. Dear God, thank You for being with us today. Keep the Holy Spirit active in our lives and in our discussions. Lead us to gain the understanding that You want for us and for all people. Hear our prayer in Jesus's name. Amen.

Rev. D: Thanks for the prayer. See you next week.

ADDENDUM 5

It's All about the Living: A Letter to the Editor

Dear Editor,

I recently ran across a small but interesting book called *Of Population and Pollution*. The author makes the argument that global warming is a consequence of overpopulation. He asserts that the ultimate consequence of doing nothing to reduce the population is the extreme likelihood that, sometime in the future, overpopulation will cause worldwide financial and social structures to fail. That, in turn, will cause the technological structure to fail, which will, finally, cause the nuclear power infrastructure to melt down and release enough radiation to bring about the extinction of all life on Earth. He claims that during that time of societal failures, there will be unimaginable harm, hardship, and destruction on Earth. Death and chaos will reign.

The author further asserts that humanity has an obligation to God to reduce the population of Earth. He asks the scientific and religious leaders of the world to work together to ensure that the population be reduced by requiring that only one child be born per couple.

If humanity does nothing, death and chaos will reign. If humanity takes on the challenge of resolving this worst-case scenario, using the ideas the author has put forward in his book, and if humanity is successful in doing so, then happiness and family solidarity will reign.

The author's message is not about death. For instance, rather than relying on abortion to end pregnancies, he asks us to accept vasectomies as the main control measure to ensure that children are not conceived. The message of this little book is about the living—humanity must keep on living. But all humans must wake up and accept their roles in resolving the overpopulation and global-warming problems that face us all.

Sincerely,
A Friend of Earth and Its People

WHY HAVE I USED A letter to the editor as a chapter in this book? Let me tell you. As I was working on this book, for quite some time I was immersed in thinking and writing about the death and destruction that would be brought on by a breakdown in society. I started to feel despondent, even jittery. I needed someone to reassure me that the reason I was writing this book was so that the living would have a better chance to continue living.

As I was considering this dilemma, I thought about you, the reader. What will you feel when you read what I have written? My next thought was, how can I reach you? I remembered the letters to the editor I had written in the past. A letter could be the means that someone might use to reassure us all. After I wrote the letter, I noticed that my despondency had gone away.

We all need to become activists and begin demanding the changes that will reduce the population and heal the atmosphere. We all need to begin writing letters to the editor. I want you to know that you have my permission to use this letter in your own way.

ADDENDUM 6

Are We at War?

IF WE ARE TO REVERSE global warming, we will need to use both hydrous and anhydrous pyrolysis in our arsenal of weapons. Did I say "arsenal of weapons?" Are we fighting a war? I think we are, in a very real way. We need a huge arsenal of hundreds of thousands of depolymerization units deployed everywhere in the world. Carbon dioxide is everywhere in the world. Plants grow everywhere in the world. We must fight this war *everywhere* in the world.

Thermal depolymerization units can be built to any size. They can even be mounted on truck trailers. Because these units are very closely related to oil refineries, the technological skills required are similar. That's why I refer to these facilities as being low tech. They can be operated by workers with journeyman-scale training. If we are to build hundreds of thousands of these facilities, we need to keep them as low tech as possible, and we may need to commandeer factories to build our arsenal, just as we did in World War II.

I remember the Apollo project, which helped the United States regain its presence in world science and leadership. A new Apollo project is needed now to perfect both hydrous and anhydrous depolymerization. Engineers and other scientists need to put their creative skills on display while they are developing units that are properly sized so that the maximum amount of petroleum and electricity can be achieved.

Another huge part of the arsenal that we will need to win this war is the infrastructure of solar panels and wind turbines that must be built at the same time that we are building the infrastructure of depolymerization units. Government at all levels must mandate that every public, business, and residential building be equipped with solar panels and that wind turbines be erected wherever feasible.

The war will be won when the world population has enough energy for transportation, heat, and light. Homes, businesses, factories, recreational facilities, and transportation can all be supplied and sustained by an infrastructure made up of depolymerization units, solar arrays, and wind turbines. Electricity from geothermal activity, tides, and wave action will also be part of our arsenal, but likely to a lesser degree.

We are spending billions or even trillions of dollars fighting over the world's reserves of oil. Then we blame religion and politics for the conflict. I say that we are spending all this money on the wrong war. It is so sad that we are able to justify spending all this money on war so that we can extract fossil fuels that should actually remain in the ground. How shamed should we feel when we see that we are using religion and politics as the scapegoats for our folly? How much better would it be to use our religious faiths, our political systems, and our money to work together and to cooperate with one another as we build a new infrastructure of alternative fuels?

This war against global warming, overpopulation, and the specter of human extinction will be won by people working together. Those whom we perceive as our enemies today will be our allies tomorrow. We will fight as soldiers together, with jobs, prosperity, and blessings for all.

ADDENDUM 7

Controlling the Message

Is it only in the United States, or is it happening elsewhere in the world? I am concerned about the balance of the political message. We all know this is important, because the rules and laws that affect our daily lives are promulgated by the people we elect to office.

The candidates who are running for office have to get their message to us so we can vote. Candidates talk to us by way of radio, television, newspapers, and electronic media. But here's the rub. The business model most commonly used in the United States and across most of the rest of the world is the capitalistic form of free enterprise. I think everyone understands that in the capitalistic model, it is natural for money to rise from the low- and middle-income earners to the capitalists at the top.

A few capitalists (oligarchs) end up with a lot of money. They can use this money to support their interests. Capitalists can own their own radio stations and other media outlets. They can get their own message out

Of Population and Pollution - A Global Warming Primer

to the people. Capitalists don't want to change what they are doing, even in the face of an impending crisis such as overpopulation and global warming. The conservative message is to keep things the same.

The liberal (progressive) message is to make changes as needed. It is apparent that the planet is being overpopulated by humans and is being overheated by pollution. The message for everyone is, "We have a problem. We see that the problem is going to get worse. We need to start doing something about it." This message will not be told on conservative talk radio because it is about change. The liberal voice in the United States and in the world needs to be heard.

So, what is happening in the genre of talk-show radio? According to ThinkProgress.org, American talk radio is dominated almost exclusively by conservative programming. In the United States, 257 talk-show stations owned by only five commercial owners air 91 percent conservative programming and 9 percent progressive programming. Also from ThinkProgress.org; 76 percent of the news/talk programming in the top 10 radio markets is conservative, while 24 percent is progressive. Conservatives have the majority of stations and listeners. Conservative talk radio has the upper hand.

Liberal talk show hosts are finding a niche by using links to the websites of some radio and television stations, but it is not enough. Liberal programming needs to be seen and heard in the airways as well. If the goal is to change the political and social makeup of our nation—including disseminating factual information about global warming and enlisting the people in doing the right thing in reducing greenhouse gasses and reducing the population—then the voices of liberals and progressives need to be as strong as (or stronger than) conservative voices.

Can we achieve equality? Yes. But we will need to build a nationwide network of progressive AM and FM radio stations, television, and print media; I also think the network should be owned by small investors. Were we to build such a network, I would name it the Global Warming Information Network, or G-WIN.

To begin this process, we will need a pool of money. For the initial pool, we will need loans from people with deep pockets. We will use this as startup money and pay it back with interest. After our network is up and running, the money to finance operations will be provided by our listeners. We will raise this money by selling subscriptions to an electronic newsletter. The cost of a subscription will become the investment amount, and it will entitle the subscriber/investor to ownership in the network. Subscription renewal will provide financial sustainability. Commercial advertising has always been a source of income for radio and other media, and commercial advertising will be part of this network, even though it will have a diminished role.

Because the organization will need to make decisions, we will form a working group to set up our network. These decisions might include making the nonprofit a cooperative or a corporation.

More about the Money

In the interests of simplification, let's refer to the financing section as Nonprofit A and the business of owning and operating the network of radio stations as Nonprofit B.

Nonprofit A will write a weekly or daily electronic newsletter that a supporter can subscribe to. Let's say that a reasonable subscription price is $6 per month ($72 per year). In addition to paying for the subscription, the $72 per year could be treated as an investment. The newsletter will be the voice that informs the reader (in the most transparent manner possible) of every aspect of the radio network, including updates on income and expenditures, radio station purchases and negotiations, interviews, and other fun facts. The newsletter will be separate from any other website.

A $72 subscription would be equal to 72 shares. The trick would be to get as many supporters as possible, each putting in $72 per year. In my particular circumstance, I have ten people in my immediate family. I would purchase ten subscriptions, at a cost of $720 per year, which would give each family member 72 shares. An individual's share value will increase year after year.

Nonprofit A will move the money to Nonprofit B by buying advertising on behalf of local, regional, or national nonprofits such as a the Humane Society, the Sierra Club, the Nature Conservancy, the ACLU, Greenpeace, a food bank, or any similar organization. Commercial advertising will also be a part of the mix, as will political advertising.

Nonprofit B will use the money to build the network and maintain operations. The goal will be to establish AM and FM stations everywhere in the United States. Operation and maintenance of each station would be under the direction of Nonprofit B. A board of directors will be chosen, employees hired, and an operations headquarters established.

MORE ABOUT THE OWNERS AND THEIR OWNERSHIP SHARES

The shareholders will be those who purchase subscriptions to the newsletter. If a subscription for $72 yields 72 shares and each subsequent renewal is added to the original, then a shareholder would have 144 shares after two years, and so forth.

Rules to administer the shares would have to be promulgated. For instance, if a shareholder dies or does not renew a subscription, those shares would be divided among the remaining shareholders. It is evident in this particular model that some shareholders will have more shares than others. This would not have to be considered a discrepancy. Rather, the shareholder with the larger number of shares could be able to cast more votes. It is also important to note that shareholders do not receive a profit from a nonprofit organization. A nonprofit will use its earnings for the expansion of its operations or for charitable purposes. How special would it be if a nonprofit, progressive media network came to be the leader in charitable donations? I think that could happen if all of its members continue to renew their yearly memberships.

I have heard of situations where a for-profit group has tried to put a nonprofit out of business by purchasing more than 50 percent of its shares. If our nonprofit is to remain viable and financially sustainable, rules will have to be made to prevent such a thing from happening.

Conclusion

The goal is to build a financially sustainable, progressive, nonprofit broadcast network that serves the entire United States with progressive programming. I think I have said enough so that you can get the gist of how I envision the business format of such a network. The final thing to say at this time is that those of you who like this idea may need to be the ones to bring this idea to fruition. I will be around, but I don't think I am the one to build a program such as I have described. It will take someone with more knowledge and experience than I have.

I am concerned that too many people think radio should be free. That is, people don't think deeply enough to understand that commercial advertising is paid for by the consumer. It will take a great deal of education and salesmanship to overcome this inertia. Sales of subscriptions will be vital for the success of this venture. It is fair to say that sales and education will have to be promoted by us all, but especially by established, progressive talk-show hosts.

ESSAY 1

In a Nutshell

WE ARE FACING THREE SERIOUS problems: (1) global warming, (2) our ever-growing piles of garbage, and (3) the need for more electricity. I believe we can solve all three of these problems if we begin using the thermal depolymerization process to change our household garbage into oil and use that oil to generate electricity (see essay 4, "Depolymerization Explained"). We must stop using fossil fuels. At the same time, we must generate electricity for our transportation needs.

Thermal depolymerization, which is not yet fully developed, is not the only process that could serve the above purpose, but I believe it is our best option. Thermal depolymerization appears to be the safest of the emerging processes, although it will likely be the most expensive to own and operate (see essay 3, "Garbage to Electricity").

I worry that large corporations and financiers—those that have the greatest monetary resources—will opt to build less expensive, less safe processes. It seems crazy, but corporations are required by law to maximize their profits.

If corporations and financiers will not build and operate thermal depolymerization plants, then the people must. I have developed a plan for a cooperative that will allow people to build depolymerization plants with investments that individuals can afford (see essay 6, "Countywide Cooperative").

Electrical generation from steam usually uses large turbines. It may be better to generate electricity by using a steam engine to drive a bank of smaller turbines, like the kinds used on wind farms (see essay 5, "Electricity by Steam Engine").

Developing small thermal depolymerization plants to change our garbage into oil, and using that oil to generate electricity, will help to solve all three of the above-named problems. Best of all, we will be able to celebrate, because we will not have to use fossil fuels in the process. In addition, since thermal depolymerization is a carbon-negative process, we will take a small amount of carbon out of our atmosphere every time we process a batch of garbage (see essay 4, "Depolymerization Explained").

ESSAY 2

The Global Dilemma

When I first wrote this essay, the carbon in the atmosphere was at 375 parts per million. Now, in 2015, it is at over *400 parts per million.* We must recite the following to everyone we speak to:

> Our scientists can measure carbon history by studying ice cores. About two hundred years ago, at the start of the industrial revolution, there were about 275 parts per million of carbon in the atmosphere. Now the measurement is about four hundred parts per million. And the polar ice is melting. Scientists tell us that we must strive to bring the carbon level back to 350 parts per million or less. If we are prudent, we will make the following rules: We must actively reduce the carbon level until we reach our goal of 350 parts per million, and we must arrange our ways of living so we will produce only as much carbon in any given year as we have removed the previous year.

This is the way that all of humanity must live their lives forever.

Let me restate this message simply: we must reduce atmospheric carbon to a safe level and then forever control how much carbon we subsequently add.

The carbon level is a measure of the amount of carbon dioxide in the atmosphere. We have an abundance of fossil fuel because at one time in the history of Earth, there was an abundance of carbon dioxide. In simple language, nature changed the carbon dioxide to coal, oil, and natural gas and then stored it underground. Of course, at that time, there weren't any humans on Earth, because mammals can't live in an atmosphere that is too high in carbon dioxide. Replacing some of the carbon dioxide with oxygen has given us the world in which we live.

Present-day scientists tell us that 350 parts per million is the highest carbon level we should maintain for the health of the planet and its occupants. The increase in carbon dioxide in the atmosphere is called the greenhouse effect, and it is warming the planet.

Global warming is forcing us to rethink the way we live. We have built our society, and everything in it, with an abundance of fossil fuel. Now, even though an abundance of fossil fuel still remains buried in the planet, we must end its use and allow that fuel to remain buried.

Let us first define the fuels we can no longer use, and then let us define the fuels that we must begin to use. We must stop using coal, crude oil, and methane gas, which nature sequestered in the Earth millions of years ago. We must begin using water, wind, wave, tidal, and geothermal activity. We must also begin using ethanol, biodiesel, oil, and methane gas from nonfossil-fuel sources. Nonfossil-fuel sources are materials such as switch grass, yellow sweet clover, hemp, algae, household garbage, agricultural waste, medical waste, human waste, and animal manure.

If we could gain all of our energy needs from non-carbon sources such as water, wind, waves, tides, and geothermal sources, we would no longer be concerned about carbon in the atmosphere. But these sources may not supply all of our energy needs. We need to put reasonable emphasis on using nonfossil-fuel sources. I suggest that we begin this process by using our household garbage.

We must determine the best way to produce energy from nonfossil-fuel, carbon-based materials. Fortunately, we have several choices. Plant matter can be changed into oil by several processes, two of which are hydrous and anhydrous pyrolysis. Other processes currently in the works require genetically engineering certain yeasts and microbes to produce ethanol and oil. Hydrous pyrolysis, also known as thermal depolymerization, as we've discussed, is the process that I believe should be our first choice. Please read essay 3, "Garbage to Electricity."

Household garbage is carbon-based. It is generally a nonfossil-fuel substance. It can be changed through thermal depolymerization into oil, gasses, fertilizer, and carbon. The oil and gasses derived through depolymerization can be used to generate electricity. This is a good thing because electricity is the energy of the future. Heating our homes, cooking our food, and fueling our cars will all be done with electricity. We must maximize our electricity production. Our garbage can be one of the fuel sources we use.

We must build thermal depolymerization plants and electrical generation plants (properly sized for each locality) together as single facilities. We have been dumping our garbage into out-of-the-way places for many years. It is now time to clean up those waste dumps.

Methane gas naturally occurs in piles of garbage, and this is a concern in waste-dump areas. Methane, as we've discussed, is a greenhouse gas that is up to twenty times more potent than carbon dioxide. Burning methane gas produces carbon dioxide. The greater danger of methane will be mitigated when it is captured and burned to generate electricity. We need to focus on using a system such as depolymerization to process the garbage in our waste dumps and thereby change the dumps back to their original natural settings.

It is good to know that thermal depolymerization plants can be built in many different sizes. We need to locate at least one appropriately sized plant in every locality, for at least five reasons:

1. A smaller thermal depolymerization plant may be more efficient.

2. Household garbage should not be hauled long distances, and existing dumps need to be cleaned up. It will be important to capture and use the methane gas at the waste dump site.
3. Every person in the United States creates four pounds of garbage every day on average. Even if we clean up all of our garbage dumps, we will always have garbage to use as raw material for depolymerization.
4. Household garbage isn't the only garbage we relegate to garbage dumps. We throw away construction debris, food production waste, and more. Local depolymerization plants will enable us to grow crops such as switch grass, yellow sweet clover, and hemp to augment garbage as the feedstock for the process.
5. Each depolymerization plant will also produce electricity. Having locally produced electricity feed into the national grid is more efficient than transporting electricity over long distances.

We must consider thermal depolymerization plants in conjunction with electrical generation plants of compatible size. In essay 5, "Electricity by Steam Engine," I spell out my ideas for using a stationary steam engine to spin generators, like the kinds used with windmills.

Spending our time discussing garbage, electricity, and thermal depolymerization is a good thing, but what good is it if no one has the money to bring these objectives to fruition? Corporations and financiers may not be interested in investing in thermal depolymerization plants because this process is likely the most expensive. Therefore, I have invented a business plan for a cooperative that could finance and operate the thermal depolymerization and electrical generation plants if corporations or private entrepreneurs will not do so. I have outlined my thoughts in essay 6, "Countywide Cooperative." In order to save our planet, we need to finance expensive projects when corporations and financiers will not do so. My idea for countywide cooperatives provides a way for the people to act when corporations and financiers will not act.

We must begin to heal our planet immediately by removing carbon from the atmosphere. We are beginning to understand that our future will be based on our ability to generate electricity without adding more carbon to the atmosphere than we can remove. The electrical-generating capacity that we in the United States now enjoy is just barely sufficient to supply everyone with enough electricity. We need to build more generating plants, but we cannot use coal, oil, natural gas, or any other kind of fossil fuel.

I talk about using fields of switch grass, agricultural debris, or other biomass to generate electricity, but it makes no sense to transport biomass a long distance to the generating plant. We need many generating plants that are smaller in size. We must develop ways to maximize the amount of electricity that is generated by smaller plants. We need one or more generating plants in every county in the United States and in every similar political jurisdiction in the world.

How much electricity can we obtain from electrical generators in conjunction with thermal depolymerization plants? I suggest that the heat, pressure, and steam that are abundant in the process will allow us to use a steam engine to drive the turbines to produce electricity.

It is useful to know that thermal depolymerization is a carbon-negative process. We must embrace this method, because it will remove carbon from the atmosphere. This is our goal.

ESSAY 3

Garbage to Electricity

THERMAL DEPOLYMERIZATION CAN TURN OUR garbage into oil; we can use that oil to generate electricity. To begin healing our planet, it is time to give companies that are involved in research and development in the thermal depolymerization process all the help they need.

Changing World Technologies (CWT) was one such company, but CWT declared Chapter 11 bankruptcy in 2009. The company has since gone out of business, and its process patent was sold to a person from Alberta who is involved with the oil industry. You can learn more about the process by Googling the company name.

We are confronting two very large problems here. One is the tremendous amount of garbage we are dumping into huge piles throughout the United States. The other is our rising need for electricity, especially as we start using electric cars. We have the opportunity to solve (or at least ameliorate) both of these problems if we can turn our garbage into electricity.

Everything I know about this issue comes from studying Internet websites. Through my research, I have learned that there are five main methods that can be employed to break garbage down into oil or other petroleum products. I will describe these five methods as concisely as possible. Please note that I may refer to garbage as "organic" or "carbonaceous" materials.

The first method is to burn the garbage and use the heat to make steam that, in turn, generates electricity. When plastic burns in the presence of oxygen, cancer-causing dioxins go up the smokestack into the environment unless the burning temperature is high enough. CaraHealth, at www.carahealth.com, in their page entitled "The Danger of Burning Plastic," reports that solid-waste incinerators can reach temperatures of up to 1,800 degrees Fahrenheit (982 C). I noticed an article from *Wikipedia* indicating that modern municipal incinerator designs show treatment for dioxins at 1,560 degrees Fahrenheit. These are no doubt acceptable temperatures to prevent the formation of dioxins. However, according to the Environment Technology Council, which is a hazardous waste resource center, the burning of toxic waste should be done at 2,200 degrees Fahrenheit. Toxic waste is different from plastic, yet I worry that toxic waste could be intermingled with plastic and processed at the wrong temperature. This method, the open-air burning of garbage, could be the cheapest method, but I shiver when I think how dangerous it might be.

A second method is anhydrous pyrolysis. Anhydrous, as noted earlier, means "without water." Pyrolysis happens when organic materials are heated to high temperatures in the absence of oxygen. Because there is no oxygen, the materials will not burst into flames; this is like making charcoal. The carbonaceous material gives off fumes that cycle through a "cracking tower" (much like one in an oil refinery) and change into oil, diesel fuel, gasoline, and natural gas, in the same way that crude oil is processed in a refinery. These fuels made from garbage can then be burned to make steam to generate electricity.

A third method is hydrous ("with water") pyrolysis. In this case, the garbage is ground up and then mixed with water and cooked under extreme heat and pressure. This process, also called thermal depolymerization, changes carbonaceous materials, such as paper and plastic, into crude oil that can be further processed in a cracking tower. This produces oil, diesel fuel, gasoline, and natural gas that can be burned to make steam to generate electricity. The thermal depolymerization process is probably the most expensive of these five processes. Each depolymerization facility will require five separate machines: the grinder, the first reactor (essentially a pressure cooker), the flash chamber, the second reactor, and the cracking tower.

A fourth method relies on genetic engineering. Yeast that has been genetically changed will feed on garbage and change it into ethanol, a form of alcohol. (To learn more about this, Google the company name Coskata,

Inc.) The ethanol formed can then be burned to make steam to generate electricity.

Be careful of genetic engineering. The yeast that nature has provided needs a lot of sugar to make alcohol, so alcohol is made only from certain plants or seeds that are high in sugar content. To be cost effective, the genetically engineered yeast will have to work on material with low sugar content. We may not want to have this kind of yeast escape into the environment. Does it make you think of an old science-fiction movie? I am sure that the folks who are working with this new yeast do not want it to escape into the environment, either. So the question that we need to ask is, what can be done to make this system fail safe?

A fifth method also relies on genetic engineering. In this method, microbes are genetically changed to feed on garbage and to excrete oil. (Begin your study here by Googling LS9, Inc.) The original microbes in nature are not very efficient at producing oil. The genetically engineered microbes are much more efficient. This process would take place in huge vats and would likely use human sewage as feedstock. The oil produced in this manner could then be burned to make steam to generate electricity. I am certain that we do not want to let genetically engineered bacteria escape into the environment, and I'm sure that the folks who are working with these new microbes also do not want them to escape into the environment. But, again, what redundancy can be built into this system to make it fail safe?

In my opinion, only one of the above methods is the safest and best. Thermal depolymerization is the best because the extreme heat and pressure, along with water, produce a magic of their own. The high temperature and high pressure will kill anything nature or genetic engineering can evolve. Microbes, yeasts, molds, mildew, and every kind of disease contained in a sample of garbage will be exterminated.

A *USA Today* article from 2004 discusses dioxins and mad cow disease. If you want more information, type "depolymerization mad cow disease dioxin" into your Internet browser. Dioxin molecules that may otherwise be released into the environment will be changed into harmless molecules when the hydrogen in the water interacts with the dioxin molecules under heat and pressure. Also, the high temperatures and pressures of this process will destroy the deadly prions (infectious proteins) associated with mad cow disease. These prions resist destruction by natural means.

Each of these methods is a part of our renewable energy future. They may all have their own niches. But even if it should cost more, the thermal depolymerization method must be the first we fully develop. Thermal depolymerization can be a fail-safe process to clean up any harm that may be caused by the escape of any genetically engineered yeast or bacteria. A fail-safe mechanism must be in place before we experiment with our well-being.

I worry that a corporation's penchant for maximizing profits will stand in the way of developing a fail-safe mechanism to protect us from unintended consequences. If private enterprise cannot or will not put thermal depolymerization into operation, then I believe that cooperatives, especially those I describe as countywide cooperatives, could be the business format of choice for this purpose. This concept is described in essay 6, "Countywide Cooperative."

ESSAY 4

Depolymerization Explained

THERMAL DEPOLYMERIZATION IS A PROCESS that breaks down solid organic substances into oil, natural gas, diesel fuel, fertilizer, carbon, and water. Organic substances include wood, paper, plastics, tires, sewage, medical waste, table scraps, garden waste, manure, common garbage, or even crops such as switch grass, yellow sweet clover, or hemp.

It is important to know the basic chemistry of organic substances. Some organic substances are very complex; they are made up of long chains of hydrocarbons called polymers. Simple organic substances are made up of shorter chains of polymers. Organic substances such as plastic, rubber, wood, and paper have long chains of polymers. Oil and natural gas are made up of shorter chains. The thermal depolymerization process uses heat and pressure to break down the long chains into shorter chains.

At the time of my original research, the aforementioned company Changing World Technologies was in the final stages of developing the depolymerization process, but it is now out of business. I do not see further development of the process for use in dealing with household garbage on an international scale. I have no way of knowing if the entity that purchased the patent process has the financial resources to put its findings to work. With any luck, the depolymerization process could be utilized in every county across the United States and in every country across the world so that we can change household garbage into usable oil and natural gas.

The machines and components that make up the mini-refinery have been available for many years. Nothing new needs to be invented to make this process work, although I expect that the initial grinding process is critical and may require further development. In other ways, the depolymerization process is closely related to refining crude oil.

Let me explain how the process works. There are five basic steps: the grinder, the first pressure cooker (often called a reactor), the depressurization chamber, another pressure cooker, and a cracking tower.

1. The grinder pulverizes the organic feedstock to the consistency of peanut butter.
2. This is mixed with water and processed at about six hundred pounds of pressure and about five hundred degrees Fahrenheit for fifteen or more minutes.
3. Next is rapid depressurization, during which the superheated water is extracted. This hot water

can be used for many purposes; one important use is as a component in producing electricity.

4. What is left after the water is gone is a slurry that is collected and put at an even higher level of heat and pressure for a longer period of time. Here, the hydrocarbons break down more completely.

5. After the slurry has gone through the second cooking process, it is transferred to a cracking tower. There it is heated again, and the fumes rise in the tower. The lightest fumes reach the highest part of the tower and become natural gas. Oil and diesel fuel come off in the middle of the tower, and carbon remains at the bottom.

6. The lightweight gasses that form at the top of the cracking tower can provide heat to run the system. This gas can also be used to generate electricity. The oil can be made into plastics, diesel fuel, or gasoline.

7. How could the carbon be used? I suppose it could be used as printing toner or as graphite. Humans have an inventive spirit; there probably is no limit to its uses. Let us remember, however, that nature has already sequestered the carbon underground as coal, petroleum oil, and natural gas. The carbon could be simply and safely sequestered in containers and stored in buildings or underground. We could use the Age of the Second Sequestration as the anthropologic name for the future.

8. The process can also extract a fertilizer that would be usable in gardens and fields.

Only 15 percent of the heat energy available in the raw materials is needed to operate the system. The components that make up a system, such as the grinding unit, the reactors, the depressurization unit, and the cracking column, can be made in any size. Think of all of these separate units being built onto one flatbed truck or being large enough to cover many acres.

Estimates vary on the amount of oil, gasses, or carbon that make up various kinds of raw material. But according to the *Wikipedia* article on thermal depolymerization, the average household garbage, when used as raw material, consists of about 40 percent oil, 20 percent lightweight gasses, 5 percent carbon, 5 percent fertilizer, and 30 percent water.

A Duke University study says that every person in the United States averages 4.3 pounds of household garbage and waste each day. That means that a county of one hundred thousand people—such as Skagit County in Washington State, where I live—can produce more than four hundred thousand pounds (two hundred tons) of garbage every day. The percentages in the previous paragraph state that household garbage will convert to 40 percent oil. If 40 percent (or 160,000 pounds / eighty tons) of the garbage is converted into light crude oil, and if a gallon of oil weighs about eight pounds, then one day's garbage from Skagit County will be converted into twenty thousand gallons of crude. Crude oil is sold by the barrel; a barrel of crude is forty-four gallons. This means that Skagit County could produce and sell about 450 barrels of oil per day. If it were sold at $40 per barrel, the county would earn about $18,000 per day, or over $60 million per year. And that's just the oil part. The amount that could be garnered from the sale of electricity and fertilizer may also be significant.

- We can expect to make money from our garbage. The final stage of the depolymerization process is to crack the crude oil into diesel fuel, gasoline, motor oil, and other petroleum products. The value of these products is higher than the cost of crude. We can burn the oil, natural gas, and diesel fuel to make electricity for our homes and the cars of the future.
- We can sell the oil to be remade into plastics.
- The diesel fuel and gasoline can be used as fuel for homes and vehicles.
- We can package the fertilizer and sell it for field and garden use.
- Probably the best we could do with the carbon that is the end result from each batch of feedstock we process will be to put it in containers and bury it in the ground. But we might be able to use a small amount for printing ink, toner, or other carbon products.

- We can also extract and sell many metals recovered from household garbage.

But of even more importance is the fact that the thermal depolymerization process is a carbon-negative process. Our goal of removing carbon from the atmosphere will be achieved.

ESSAY 5

Electricity by Steam Engine

WE CAN DISPOSE OF HOUSEHOLD garbage by using the process of thermal depolymerization to change our garbage to oil. Next, we can use that oil to make steam to generate electricity. A facility to dispose of our garbage would consist of both a mini-refinery and an electrical generation plant.

As kids growing up in the northern Midwest, my buddies and I would run down to the town depot when we spotted the train coming around the bend into our small town. (Yes, we actually had a big bend in the tracks about half a mile south of our town.) The train would come chugging in with its smokestack belching. After the train stopped, clouds of steam would hiss as the steam engine depressurized.

Those locomotives could do a lot of work. I can envision how a steam engine can be put to work to generate electricity. I want to bring this vision to you.

First, let me describe the way electricity is usually generated. Water is boiled in a closed chamber until the steam is under pressure. A valve is opened, and the steam is directed at fins on a cylindrical rotor. This causes the rotor to spin. The rotor spins through a magnetic field, generating electricity.

Large generating plants can be located hundreds of miles from the cities that use the electricity. Hydropower, nuclear fission, or coal-fired plants have huge rotors. The problem with using large rotors to supply electricity is that the rotors must generate electricity twenty-four hours a day, and much of it goes unused. You have heard that electric companies will sell electricity at a cheaper rate during off-peak hours. This is because the rotors at the generating plant have not been turned off. Electricity that is not used is wasted. It makes sense to use smaller generators that can be turned on or off at will.

Smaller generators, like those used in windmills, can be powered by steam engines as well as by windmill blades. A steam engine turns a flywheel. The circular motion of the flywheel can drive a chain attached to a bank of windmill generators of various sizes. One steam engine could drive more than one bank of generators. I am thinking that having two or more stationary steam engines at each depolymerization site, each driving one or more banks of windmill generators, would provide all the flexibility needed. We could even design a computer program to turn the generators on or off, thus controlling the output and maximizing efficiency. Very little, if any, electricity will be wasted, because the generation plants will produce only the amount of electricity needed at any given time. Increasing our use of these generators will not only be a boon to the windmill generator industry but will produce many new jobs.

Where should we build our depolymerization/electrical-generation plants? Should several states join together and build a megasized plant for a region? Should we build one large plant in each state? Should a smaller plant be built for every county? Perhaps every county should have several plants? I suppose it depends on how far we want to haul our garbage.

In addition, certain agricultural crops could be grown to augment the amount of garbage available. As I have noted earlier, crops such as switch grass, yellow sweet clover, and hemp all appear to be viable candidates. Clover is especially good because it is a legume that naturally fertilizes the soil by adding nitrogen. It makes sense to grow these crops close to the depolymerization/electrical-generation sites.

What is the relative cost of building one large plant to that of building several small plants that would equal the output of one large plant? This needs to be confirmed, but I think that building one large plant would cost less. Further, I believe that large corporations or financiers may stay away from building small plants because of the higher costs involved. Even so, small thermal depolymerization/electrical-generation plants scattered around a county may save money and time in transportation costs. In addition, both of these processes use heat and pressure, which can be dangerous. I think the adage "smaller is better" finds sensible application with these processes.

I worry that private businesses or large corporations will be interested only in profits. They will remove the profits from the local area and try to do everything cheaply. How much better would it be to use local nonprofits to build the infrastructure to meet the needs of the people?

Even though building several small plants may cost more money than building a large generating plant, I believe that building many smaller plants close to the sources of the raw materials will cost less in the long run. This is where the idea for a countywide cooperative comes into play. A countywide cooperative is a business plan that allows every person who lives in a locality to invest in a cooperative at very little cost. The cooperative business plan I envision will be local. Profits will be spent or reinvested locally. I write about my idea for a countywide cooperative in the next essay.

ESSAY 6

Countywide Cooperative

We generally think of a "cooperative" in terms of a group of people who have a similar interest. A group of farmers might join together as a cooperative to build a processing plant for the crops they grow; each farmer has an ownership in the cooperative.

Ownership is usually defined in terms of shares. In some cooperatives, every owner puts an equal amount of money into the cooperative, thereby establishing equal shares. In other cooperatives, some owners put in more money than others and are entitled to larger shares of the profits.

I will use this essay to talk about a cooperative of my own invention that I choose to call a countywide cooperative. My intention is that all of the people in a county have equal shares in a cooperative that changes their household garbage into energy.

In the previous essay, entitled "Garbage into Electricity," I pointed out five emerging methods that could turn our household garbage into electricity. Of these, the process called thermal depolymerization is likely the safest, but it is also likely the most expensive to use. I also expressed my concern that thermal depolymerization facilities may never be built because corporations and financiers will choose to invest in cheaper methods even though those may be less safe. The corporations may not invest in the more expensive process because they may not be able to maximize their profits.

I have devised the idea of a countywide cooperative in an effort to give people like me (the common, middle-class people of a county) a chance to join together to build and operate a facility that uses the thermal depolymerization process to generate electricity from their own garbage. As the name implies, a countywide cooperative would require ownership by every person (man, woman, and child) in a county. Each person would have an equal share. Every person, no matter how young or old, generates garbage. Every person is a stakeholder in dealing with our garbage.

Because we need to stop dumping our garbage into huge piles, we need one or more thermal depolymerization plants in every county (depending on the size, population, and other characteristics of the county) in the United States. A thermal depolymerization plant can be built to any size. It can be huge, covering many acres, or it can be a unit that could sit on a flatbed truck. It may be to our advantage to have two or more smaller plants located within the county, rather than one large plant.

The only information I could find about the cost of a plant is in a reference to a grant from the EPA to CWT for about $15 million to build four plants the size of the experimental plant the firm had built in Philadelphia. (As mentioned, the company is now defunct, and the process patent has been sold.)

For the purpose of discussing my idea for a business plan, I need to have a value for a plant of usable size. Let me pluck a value of $24 million from the air. Assuming that an experimental plant would be smaller than a usable plant, a value of $24 million may be a reasonable figure. Let us process that $24 million through my business plan to see how it works out. In real life, the value that is used might be larger or smaller than that, depending on the size and other factors.

As I mentioned, I live in Skagit County, north of Seattle, a large, mostly rural county of about 120,000 people. Skagit County would have to raise $24 million to establish our first plant. The recognized procedure for county governments to raise money for infrastructure such as roads, bridges, and buildings is to issue and sell municipal bonds with a promise to repay the bond with interest over a twenty- or thirty-year period. The people who buy these bonds do not have to live in Skagit County. They can live anywhere in the world. The county has to make good on the bonds, regardless of profit. Under this system, the bond owners couldn't care less about the county or the local project. They are interested only in the profits from their investment.

Because it seems to me that a more positive arrangement to raise the necessary money can be devised, I have conceived of a more positive arrangement: the county-wide cooperative. I mentioned that the population of Skagit County is about 120,000 people. If every man, woman, and child in the county invested $200 in a cooperative in Skagit County, we could raise $24 million. The cooperative would be owned by all the people who reside in the county. No one would be left out.

A naysayer will immediately point out that not everyone will be able to afford—or be inclined—to buy a share for $200. This can be overcome as follows:

* A Good Samaritan could pay for someone who either can't or won't buy a share and would be repaid from the earnings the other person would have received had that person purchased the share. A reasonable rate of interest could be added to encourage the Good Samaritan. For instance, if I put up $20,000 for shares, I could buy one for myself, my wife, and one for each of ninety-eight other county residents. The money to pay me back (with interest) would come out of the earnings of these other people's shares until I am fully reimbursed, whereupon the share would be signed over to the resident. The goal would be to make certain that every man, woman, and child becomes a shareholder even if some people become shareholders by virtue of help from others.
* Philanthropists, governments, and civic organizations could be allowed to provide the money

needed to make certain that everyone in the county becomes an owner.

In addition to making every person a shareholder at the time the cooperative is started, the following will apply:

- When people are born in (or move into) the county, they will be required to pay $200 each into the cooperative to establish a share of ownership. If the newcomers do not have the money, they may have to depend on a Good Samaritan, a service club, or some other organization that has begun a fund for that purpose.
- When people die or move out of the county, their accounts will be closed. The people who move out will be paid the amount of the original investment ($200), plus any amount over that $200 that remains in the account. If the value of the account is less than $200, then the only payment would be the original investment of $200. In the case of death, that person's heirs will be paid in the same manner as someone who moves out of the county.

The cooperative will have a business office in the county. Clerks will be on hand to deal with opening accounts for new owners and closing accounts when owners die or move.

No person would be allowed to own more than one share. This means that shares will not be sold between shareholders. No entity that is not a live, physical person will be allowed to own any shares in the cooperative. Individual accounts will increase and decrease as follows:

- In the beginning, an account will have $200 because that is the initial investment.
- The first thing that will happen will be to spend the money needed to build the plant and establish the business. The account will decrease in value during this time.
- As the plant begins operation, profits will increase the value of the account. A dividend of $200 will be paid to the owner when the value of the account reaches $400.
- Dividends could be paid on the purchase anniversary, as long as the account contains at least $400.
- It may be in the best interest of the cooperative to allow the owners' accounts to exceed $400. Extra money can always be used for related projects such as solar panels or windmills.
- No dividends will be paid to an address outside of the county.
- An accounting system will be formulated to distribute all of the income and expenditures evenly among the shareholders all of the time.

Will the residents of a county actually make any money off their $200 investment? Yes, they will. But does it really make a difference if the residents do not make money from their investment? After all, a payment of only $200, made only once in a lifetime, knowing that it will be repaid to your heirs, is more like an act of volunteerism than an investment.

How can money be made by turning garbage into electricity? Using my estimates from earlier, the thermal depolymerization process will change garbage to 40 percent oil, 20 percent lightweight gasses, 5 percent carbon, 5 percent fertilizer, and 30 percent sterilized water. This means that a ton of garbage will provide eight hundred pounds (roughly one hundred gallons) of refined oil. One hundred thousand people will produce about two hundred tons of garbage per day. That calculates to twenty thousand gallons of oil per day. At $5 per gallon, that equals $100,000 per day.

The lightweight gasses (such as methane) can be burned to make electricity. The oil can be burned to make electricity or sold to be remanufactured into plastics or other petroleum products that can then be sold. The fertilizer can be sold for farm and garden use. This doesn't even take into account all of the metal that will be sorted out of the garbage. Rest assured, there will be a profit.

But keep in mind, if there are 120,000 shareholders, it will take $24 million of accumulated profit to make a dividend payment of $200. The first dividend might not be paid in the first year. The reality is that the $200 original investment is such a small amount that it cannot be regarded as an investment for profit. It is an investment to improve our lives and our planet. Any profit earned will be like icing on the cake.

Each cooperative will have to elect a board of directors and hire a manager and other workers. These people will need to be residents of the county. An honest board will oversee the honest, hard work of the manager and employees. (As an aside, I recommend that the highest-paid earner in the cooperative should not be paid more than four times the amount paid to the lowest-paid person.)

Creating this countywide cooperative will enable us to solve, or at least ameliorate, our problems with household garbage by turning it into electricity. It is in our best interest to use our garbage rather than pile it higher and higher. The ideas I have put together to describe the cooperative business arrangement that I call a countywide cooperative are the result of my own thought processes. I regard these ideas as my own intellectual property, but even as I claim my intellectual property rights, I want this idea to be spread across the world as quickly as possible. The fate of the planet may rest on this single idea.

About the Author

Who am I to be so audacious as to tell the world's religious, scientific, government, and business leaders what they are to do and how they are to solve the global problems we now face? Well, someone needs to kick-start an effort to solve the overpopulated, overpolluted, and overnuclearized situation the world is currently in. Someone needs to be the first voice to say that leaders in science must convince religious leaders that overpopulation will, at some point, bring down our civilization. Someone needs to be the first voice to say that religious leaders must then convince government and business leaders that there truly is a specter of annihilation looming over humanity and that government and business need to put the world's money and resources to use in building an infrastructure of hundreds of thousands of depolymerization units. Someone needs to be the first voice to tell the common folk to look to their religious leaders for strength and support in their extremely important role of reducing the population by having only one child. And someone needs to tell homeowners to be willing to comply with the idea of using their lawns as places to grow our food.

I'm not a scientist. I'm not a journalist. I'm not known as an expert on anything. I am retired and a member of the middle class. I have been looking at and thinking about the ramifications of global warming for quite some time, and I came up with the theories and solutions presented here as a way to confront the issues of overpopulation, global warming, and nuclear meltdown. I have a bachelor's degree (1963) from Mayville State, a small college in North Dakota, where I majored in mathematics and business education. I began my work life as a teacher but retired from a career as a US customs inspector. After retiring, I became involved in drug-policy reform, and I now belong to a group called Law Enforcement Against Prohibition. I also worked with the King County (Washington) Bar Association Drug Policy Reform Task Force. You can find my biography and a picture at www.leap.cc under the "Find a Speaker" tab.

I have written this primer explaining the worst-case scenario relative to overpopulation, global warming and nuclear power plants. Now it is up to someone else to write about the ravages of a thermohaline circulation shutdown or the loss of resources such as forests, fisheries, soil, water, phosphorous, copper, rare-earth elements and more. It is sad—but, I suppose, only realistic—that overpopulation is creating worst-case scenarios everywhere. In the future our progeny must live differently than humanity has lived, since the beginning of time.

Help them do so. Know that we, the generations living today, are the people who must make the transition to this new paradigm.

www.ingramcontent.com/pod-product-compliance
Lightning Source LLC
Chambersburg PA
CBHW051908170526
45168CB00001B/291